B7/K

1550 1600 10 -1650

AF699898

TYCHO BRAHE 1446-1601

546

JOHANNES KEPLER 1571-1630

-1550

G RHETICUS 1514-1576

ER 1498-1552

547

GALILEO GALILEI 1564-1642

D 1511-1553

RENE DESCARTES 1596-1650

543

CARDANO 1501-1576

MICHAEL MAESTLIN 1550-1631

GIORDANO BRUNO 1548-1600

Copernicus

Copernicus

VISTAS IN ASTRONOMY VOLUME 17

Yesterday and Today

Proceedings of the Commemorative Conference held in Washington in honour of Nicolaus Copernicus

Edited by
ARTHUR BEER AND K. AA. STRAND

PERGAMON PRESS
Oxford · New York · Toronto · Sydney · Braunschweig

Pergamon Press Ltd., Headington Hill Hall, Oxford
Pergamon Press Inc., Maxwell House, Fairview Park, Elmsford, New York 10523
Pergamon of Canada Ltd., 207 Queen's Quay West, Toronto 1
Pergamon Press (Aust.) Pty. Ltd., 19a Boundary Street,
Rushcutters Bay, N.S.W. 2011, Australia
Pergamon Press GmbH, Burgplatz 1, Braunschweig 3300, West Germany

---

Copyright © 1975 Pergamon Press Ltd.

All Rights Reserved. No part of this publication may be reproduced, stored in a retrieval system, or transmitted, in any form or by any means, electronic, mechanical, photocopying, recording or otherwise, without the prior permission of Pergamon Press Ltd.

First Edition 1975

Library of Congress Catalog Card No. 56-4213

0 08 017878 2

Film prepared by
Biddles Ltd, Guildford, Surrey
Printed in Great Britain by
Butler & Tanner Ltd,
Frome, Somerset

Contents

Foreword

"... The rest of the story is quickly told. The book appeared in March 1543, but a copy did not reach Copernicus until May. He was then on his deathbed, and his actual death occurred within hours of the book's arrival.

"To the citizens of Frauenburg it must only have seemed as if an infirm old man of some past distinction had passed from their midst. They were not to know that drums were already rolling with the noise of distant thunder. His fellow-canons, as they laid him in his grave, were not to know that Copernicus had detonated an overwhelming explosion of human knowledge, an explosion still with us in our own day, and whose eventual outcome we cannot yet foresee."

From FRED HOYLE, *Nicolaus Copernicus. An Essay on His Life and Work.*
Heinemann, London, 1973.

By way of Introduction

Institute of Astronomy, University of Cambridge, England — ARTHUR BEER

U.S. Naval Observatory, Washington, D.C. — K. AA. STRAND

Preface

On December 27 and 28, 1972, a conference was held in Washington, D.C., to commemorate the 500th anniversary of the birth of Nicolaus Copernicus in Toruń, Poland, on February 19, 1473.

The Conference was held to coincide with the annual meeting of the American Association for the Advancement of Science.

A group of forty-five scholars from Europe, Canada and the United States were selected by a U.S. National Advisory Committee and invited to participate in the Conference to discuss the scientific and humanistic significance of our Copernican heritage.

The Conference was co-sponsored by the U.S. Naval Observatory, the National Graduate University of Washington, D.C., and by the American Association for the Advancement of Science. The National Science Foundation gave financial support, which is gratefully acknowledged.

On behalf of the Advisory Committee, it is a pleasure to thank the contributors of the invited papers, those who participated in the discussions, and especially the chairmen of the sessions for their efforts in organizing their programs.

It is a pleasure to acknowledge the generosity of the contributors of the many interesting illustrations. Particularly, in this respect, I am grateful to Charles Eames of Venice, California, and Mrs. Jeannine Oppenwall of his staff for her courteous and prompt response to my many requests, and to Bern Dibner of the Burndy Library of Norwalk, Connecticut.

I am especially thankful to the members of the staff of the U.S. Naval Observatory: P. Kenneth Seidelmann and Paul M. Janiczek for handling the recording of the discussions taking place at the Conference, Mrs. Ruth L. Meyers for transcribing them, Mrs. Josephine Lanterman for handling the extensive correspondence and her assistance in preparing the manuscripts, Varkey V. Kallarakal for reading proofs, and Mihran Miranian, who did the photographic work of copying the illustrations from the Burndy Library, and the library of the U.S. Naval Observatory.

Thanks are due in particular to Dr. Arthur Beer, of the University of Cambridge, England, and editor of *Vistas in Astronomy*, for so generously providing for the publication of the Proceedings of this Conference in a special volume of this series.

Washington, D.C.

K. AA. STRAND

Advisory Committee

JEAN K. BOEK, National Graduate University, Washington, D.C.

STILLMAN DRAKE, University of Toronto, Toronto, Canada.

RAYNOR L. DUNCOMBE, U.S. Naval Observatory, Washington, D.C.

OWEN J. GINGERICH, Harvard College Observatory, Cambridge, Massachusetts.

C. DORIS HELLMAN, Queens College, City University of New York, N.Y.

ARLO U. LANDOLT, Louisiana State University, Baton Rouge, Louisiana.

ERNEST NAGEL, Columbia University, New York, N.Y.

EDWARD ROSEN, City University of New York, New York, N.Y.

RAYMOND J. SEEGER, American Association for the Advancement of Science, Washington, D.C.

K. AA. STRAND (*Chairman*), U.S. Naval Observatory, Washington, D.C.

ANTONI ZYGMUND, University of Chicago, Chicago, Illinois.

Session Chairmen

OWEN GINGERICH:	1
EDWARD ROSEN:	2
ERNEST NAGEL:	3
WALTER E. BOEK:	4
K. AA. STRAND:	5

Participants

SILVIO A. BEDINI, National Museum of History and Technology of the Smithsonian Institution, Washington, D.C.

ARTHUR BEER, University of Cambridge, Cambridge, England.

PETER BEER, Science Unit, British Broadcasting Corporation, London, England.

RICHARD E. BERENDZEN, Boston University, Boston, Massachusetts.

JEAN K. BOEK, National Graduate University, Washington, D.C.

WALTER E. BOEK, National Graduate University, Washington, D.C.

CARL B. BOYER, Brooklyn College, New York, New York.

STEPHEN BRUSH, University of Maryland, College Park, Maryland.

ALISTAIR C. CROMBIE, All Souls College, Oxford University, Oxford, England.

JOHN M. DANBY, North Carolina State University, Raleigh, North Carolina.

RADM T. F. DEDMAN, USN, Department of the Navy, Washington, D.C.

JOHN DILLENBERGER, Graduate Theological Union, Berkeley, California.

JERZY DOBRZYCKI, Polish Academy of Sciences, Warsaw, Poland.

STILLMAN DRAKE, Institute for the History and Philosophy of Science and Technology, University of Toronto, Toronto, Canada.

RAYNOR L. DUNCOMBE, U.S. Naval Observatory, Washington, D.C.

LANGDON GILKEY, University of Chicago, Chicago, Illinois.

OWEN GINGERICH, Harvard University, Cambridge, Massachusetts.

WULFF D. HEINTZ, Swarthmore College, Pennsylvania.

DORIS HELLMAN, Queens College, City University of New York, New York, New York.

PAUL HERGET, Cincinnati Observatory, Cincinnati, Ohio.

GERALD J. HOLTON, Harvard University, Cambridge, Massachusetts.

WILHELMINA IWANOWSKA, Astronomical Observatory, Toruń, Poland.

MARTIN J. KLEIN, Yale University, New Haven, Connecticut.

ARLO U. LANDOLT, Louisiana State University, Baton Rouge, Louisiana.

Participants

ROBERT B. LINDSAY, Brown University, Providence, Rhode Island.
ERNEST NAGEL, Columbia University, New York, New York.
SHIGERU NAKAYAMA, Harvard University, Cambridge, Massachusetts.
BENJAMIN NELSON, New School for Social Research, New York, New York.
JOHN NORTH, Museum for the History of Science, Oxford University, Oxford, England.
ROBERT PALTER, University of Texas, Austin, Texas.
JEROME RAYMOND RAVETZ, Leeds University, Leeds, England.
MARIA LUISA BONELLI RIGHINI, Institute and Museum of History of Science, Florence, Italy.
P. L. ROSE, New York University, New York, New York.
EDWARD ROSEN, City University of New York, New York, New York.
RAYMOND J. SEEGER, National Graduate University, Washington, D.C.
THOMAS B. SETTLE, Polytechnic Institute of New York, New York, New York.
DUDLEY SHAPERE, University of Illinois, Urbana, Illinois.
WILLIAM R. SHEA, University of Ottawa, Ottawa, Canada.
JOVES SOLDACZUK, Polish Embassy, Washington, D.C.
K. AA. STRAND, U.S. Naval Observatory, Washington, D.C.
NOEL SWERDLOW, University of Chicago, Chicago, Illinois.
VICTOR SZEBEHELY, University of Texas, Austin, Texas.
STEPHEN E. TOULMIN, University of California, Santa Cruz, California.
GART WESTERHOUT, University of Maryland, College Park, Maryland.

Introductory Remarks

RADM T. F. DEDMAN, USN

Asst. Vice Chief of Naval Operations

DR. STRAND, Captain Davidson, members and guests of the American Association for the Advancement of Science. I am delighted to welcome you all to the Naval Observatory in the name of our Secretary of the Navy, the Honorable John Warner, and our Chief of Naval Operations, Admiral Zumwalt.

It seems especially fitting that the United States Navy, with its rich scientific heritage in astronomy as represented by the U.S. Naval Observatory, should host such a distinguished group of international scholars as yourselves as you gather to commemorate the 500th anniversary of the birth of Copernicus.

Your program over the next two days, and the eminent backgrounds of the philosophers, historians, and scientists gathered here today, reflect the many-faceted career of this brilliant Renaissance scholar who is now recognized as the "Father of Modern Astronomy".

To you gentlemen, who measure the distances by the speed of light, the 500 years since the birth of Copernicus represents a relatively short span of time. However, it is interesting to speculate on whether discoveries and scientific achievements being made today will have any greater impact on our world than did the discoveries of Copernicus on the world of the sixteenth century.

A little over a week ago, a United States Naval Officer, Captain Eugene Cernan, commented while exploring the Taurus–Littrow region of the surface of the Moon, in company with his civilian counterpart, Dr. Schmitt, that the Apollo missions had changed both mankind and the world as we know it. Our hosts today, the staff of the U.S. Naval Observatory, can take great pride in the programs of research in astrometry, and celestial navigation and mechanics which contributed so much to the success of the Apollo program and thus to the changes in the world being brought about by that program. Captain Cernan and Dr. Schmitt navigated the lunar module "Challenger" to within 100 meters of the planned landing point at "Camelot" Crater. This remarkable feat of navigation, like the exploits of all earlier Apollo astronauts, was made possible by special observations, computations and tables prepared here at the Naval Observatory. I am confident that the Naval Observatory will continue to show the way for our space travelers in future missions, both in our own national programs, and also in those programs in which we join with other nations, as Copernicus showed us all the way to understand our heliocentric system nearly 500 years ago.

Again, welcome aboard the Naval Observatory, and on behalf of Admiral Zumwalt, I wish you a most stimulating and profitable conference.

Editorial Note

Relevant book-pages from the fifteenth and sixteenth centuries relating to our Copernican themes (Figs. 1–25) are reproduced on pp. xviii–xxxiii.

Nicolaus Copernicus (*1473–1543*)

No portrait known to have been painted during his lifetime is extant; the above reproduction refers to an oil painting on wood panel from the mid-sixteenth century. It seems to represent an authentic tradition of Copernicus' appearance, agreeing reasonably with a portrait (now preserved in Strasbourg) by the German artist Tobias Stimmer, who it is said copied from a now lost self-portrait of Copernicus. (Reproduced by permission of the Municipal Museum, Toruń, Poland, from a colour photograph taken by Charles Eames in 1972.)

FIG. 1. Copernicus in Warsaw. Photograph, by courtesy of Klaus J. Kallabis. (*Die Zeit*, Hamburg.)

FIG. 1A. Copernicus' task is well illustrated by this photograph taken from the artificial sky of the Planetarium in Munich (Deutsches Museum, München): the motors of its drive were adjusted to give a superposition of planetary motions covering several years. This geocentric view of the Universe was transformed by Copernicus.

FIG. 2. The autograph manuscript of *De revolutionibus*, today preserved in the Library of the Jagiellonian University in Krakow. (Photograph by Charles Eames.)

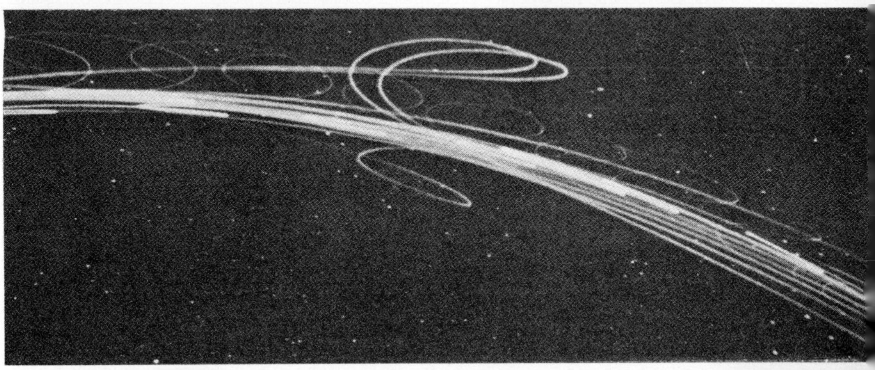

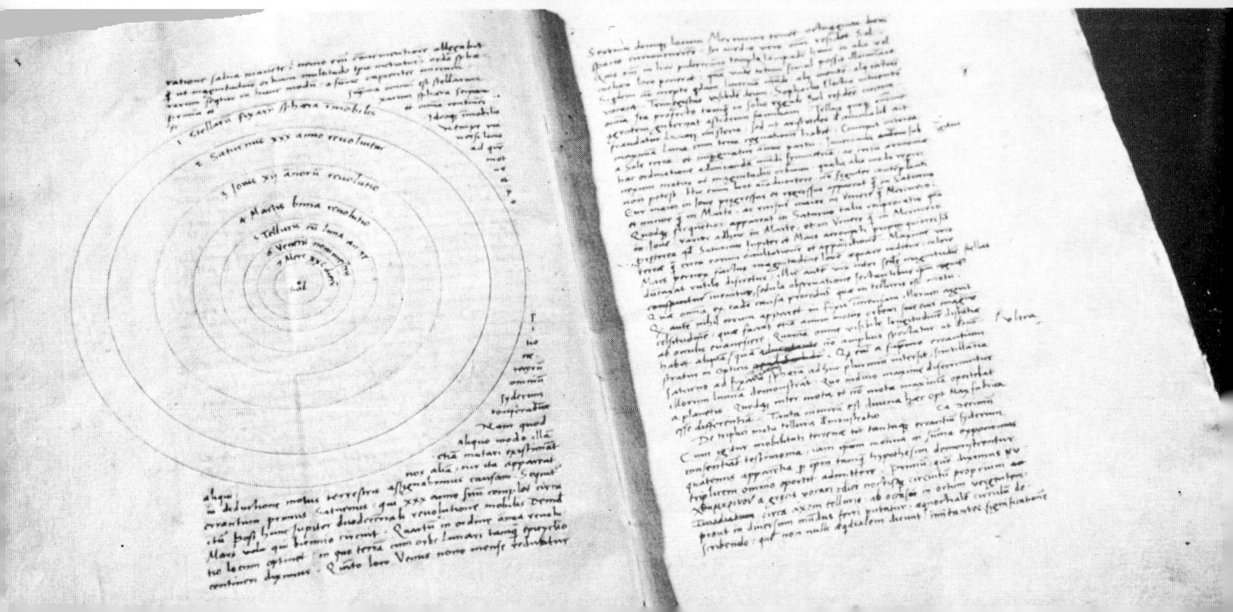

NICOLAI COPERNICI

net, in quo terram cum orbe lunari tanquam epicyclo contineri diximus. Quinto loco Venus nono menſe reducitur. Sextum deniq; locum Mercurius tenet, octuaginta dierum ſpacio circũ currens. In medio uero omnium reſidet Sol. Quis enim in hoc

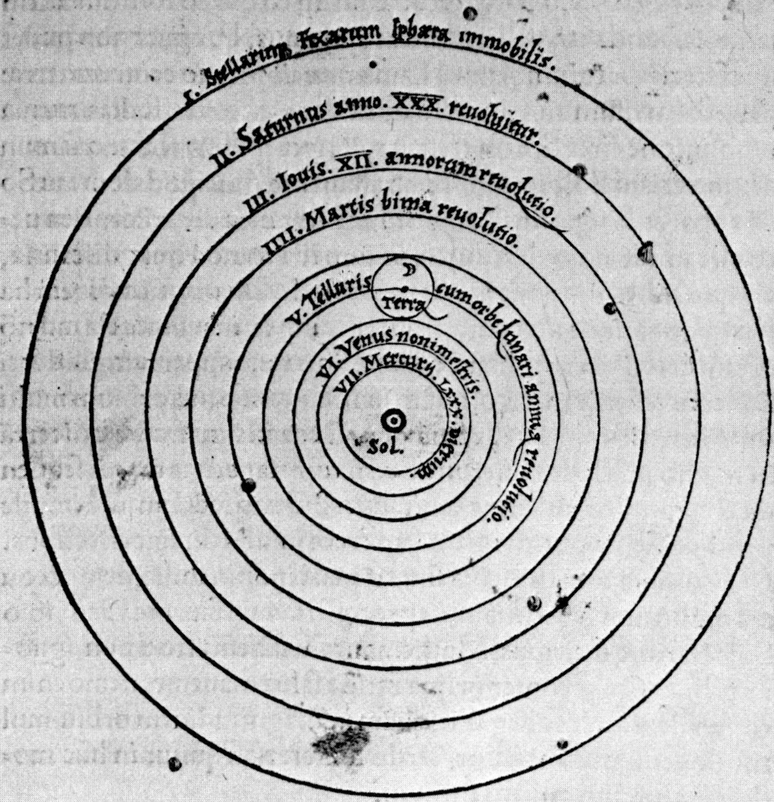

pulcherimo templo lampadem hanc in alio uel meliori loco poneret, quàm unde totum ſimul poſsit illuminare? Siquidem non inepte quidam lucernam mundi, alij mentem, alij rectorem uocant. Trimegiſtus uiſibilem Deum, Sophoclis Electra intuentẽ omnia. Ita profecto tanquam in ſolio regali Sol reſidens circum agentem gubernat Aſtrorum familiam. Tellus quoq; minime fraudatur lunari miniſterio, ſed ut Ariſtoteles de animalibus ait, maximã Luna cũ terra cognationẽ habet. Concipit interea à Sole terra, & impregnatur annuo partu. Inuenimus igitur ſub hac

FIG. 3. The left-hand page of Fig. 2 (manuscript) in its printed form. (Photograph by Owen Gingerich. Permission of Harvard College Library.)

FIG. 4. The autograph manuscript of *De revolutionibus*, today preserved in the Library of the Jagiellonian University in Krakow. (Photograph by Charles Eames.)

FIG. 5. Nicolaus Copernicus. From P. Gassendi, *N. Copernici Vita*, 1654. Engraving, seventeenth century. (Courtesy of Burndy Library.)

FIG. 6. Nicolaus Copernicus. J. J. Boissard, Icones, 1957. (Courtesy of Burndy Library.)

FIG. 7. Nicolaus Copernicus. Engraving, seventeenth century, Paris. (Courtesy of Burndy Library.)

Fig. 8. Personification of Astronomy, from the *Margarita Philosophica* of Gregor Reisch. (Photograph by Charles Eames.)

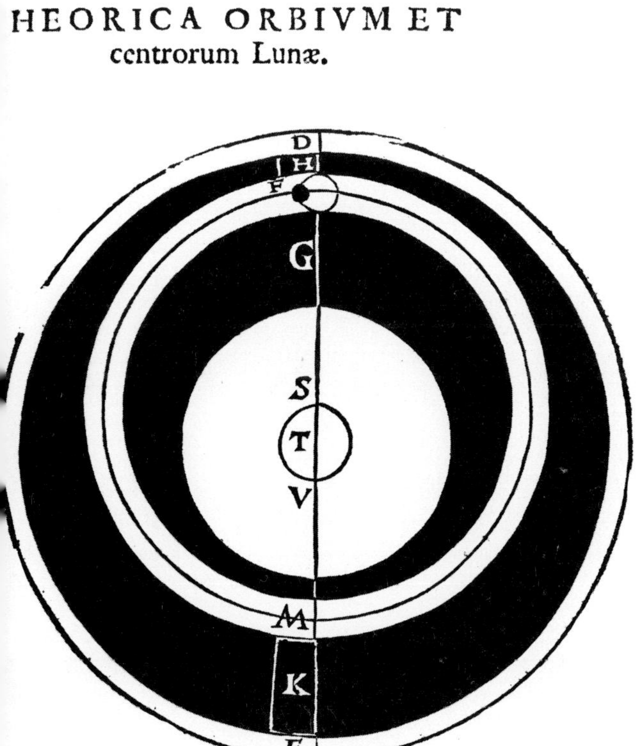

Fig. 9. Illustration of the lunar orbit in George Peurbach's *Theoricae Novae Planetarum* in the 1556 edition. First published in 1473. (U.S. Naval Observatory Library.)

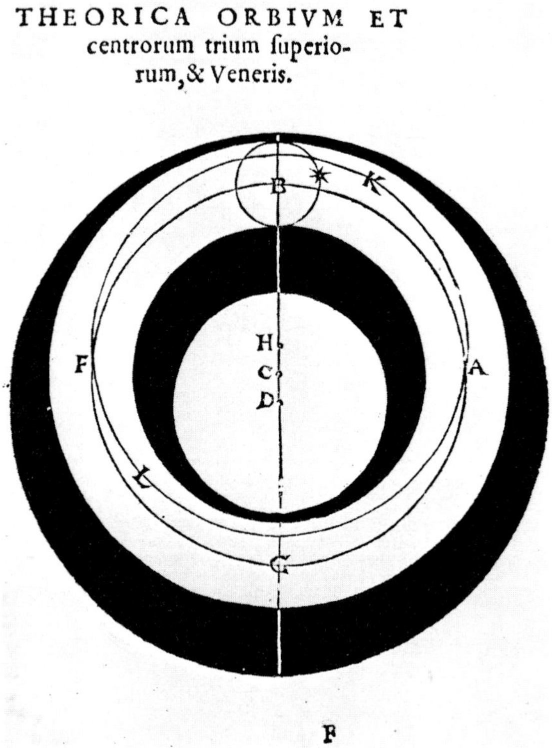

FIG. 10. Theoretical Orbit of Venus in George Peurbach's *Theoricae Novae Planetarum* in the 1556 edition. First published in 1473. (U.S. Naval Observatory Library.)

FIG. 11. The epicyclic theory in Leopoldus' *Compilatio de Astrorum Scientia* (Augsburg, Ratdolt, 1489). (From the collection of Owen Gingerich. Photograph by Charles Eames.)

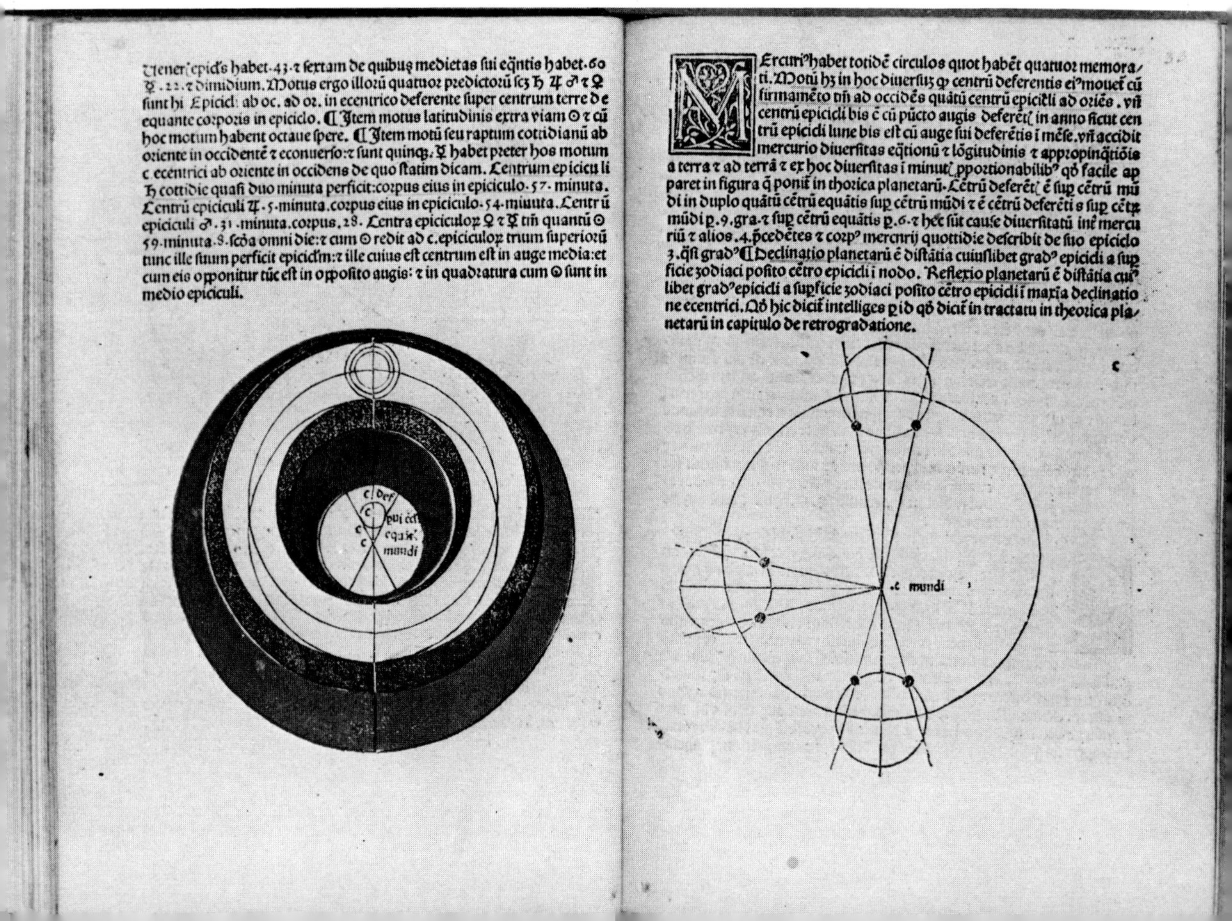

FIG. 12. An Aristotelian cosmology, from Hartmann Schedel, *Nuremberg Chronicle* (1493). (From the collection of Owen Gingerich. Photograph by Charles Eames.)

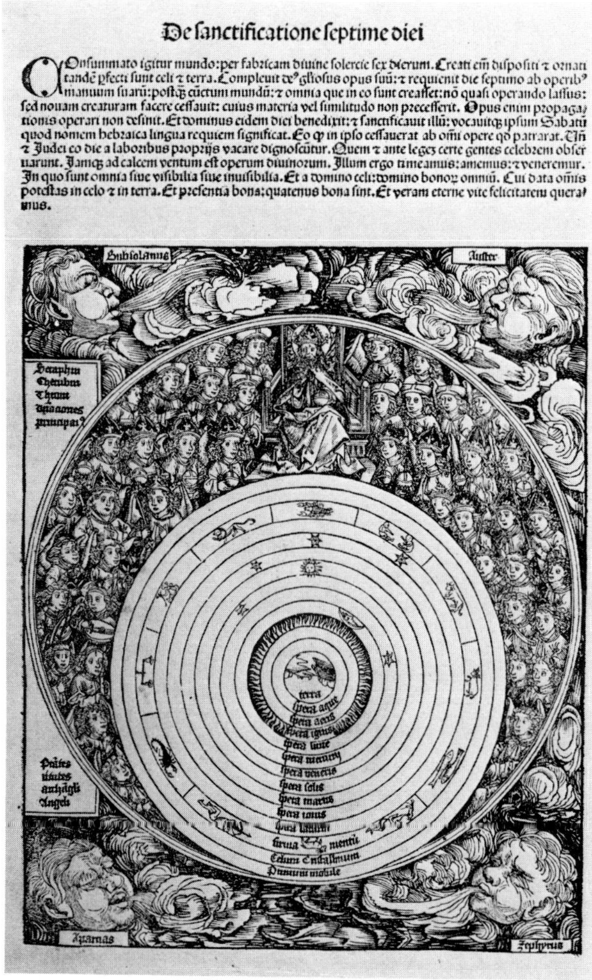

De sanctificatione septime diei

Consummato igitur mundo: per fabricam diuine solercie sex dierum. Creati eni dispositi & ornati tandē pfecti sunt celi & terra. Compleuit de' gliosus opus suū: & requieuit die septimo ab operib' manuum suarū: postq cūctum mundū: & omnia que in eo sunt creasset: nō quasi operando lassus: sed nouam creaturam facere cessauit: cuius materia vel similitudo non precesserit. Opus enim propagationis operari non desinit. Et dominus eidem diei benedixit: & sanctificauit illū: vocauitq ipsum Sabatū quod nomem hebraica lingua requiem significat. Eo q in ipso cessauerat ab omi opere qd patrarat. Vñ & Iudei eo die a laboribus proprijs vacare dignoscūtur. Quem & ante leges certe gentes celebrem obseruarunt. Iamq ad calcem ventum est operum diuinorum. Illum ergo timeamus: amemus: & veneremur. In quo sunt omnia siue visibilia siue inuisibilia. Et a domino celi: domino bonoz omniū. Cui data omis potestas in celo & in terra. Et presentia bona: quatenus bona sint. Et veram eterne vite felicitatem queramus.

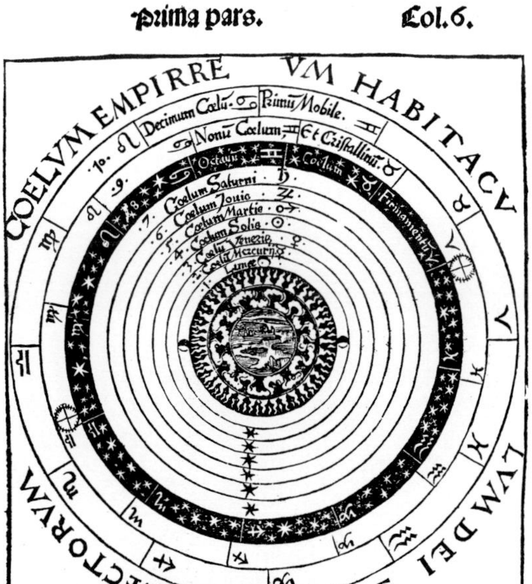

FIG. 13. An Aristotelian cosmology from Peter Apian's *Cosmographia* (1524). First edition. (Photograph by Charles Eames.)

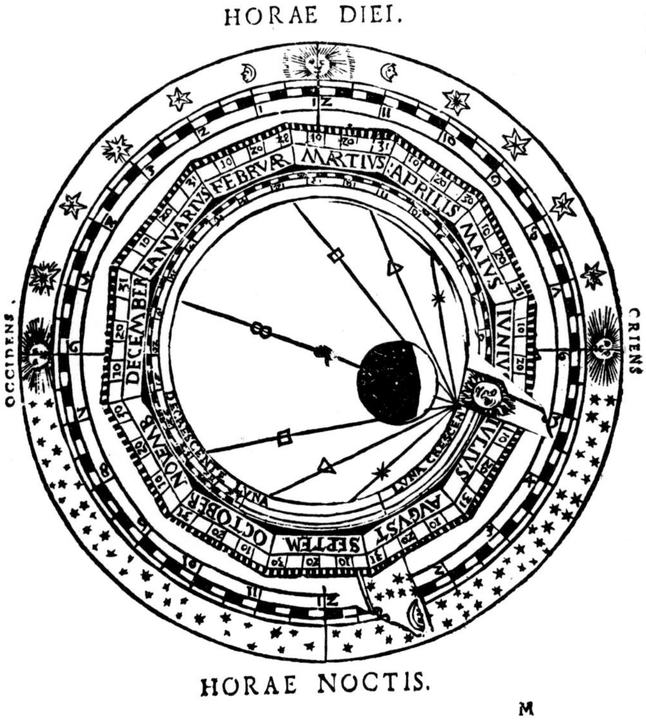

FIG. 14. A volvelle from Peter Apian's *Cosmographia* (1540), for determining the phases of the Moon. (U.S. Naval Observatory Library.)

FIG. 15. From Peter Apian's *Cosmographia* (1540): "The Earth surrounded by the twelve winds."
(U.S. Naval Observatory Library.)

FIG. 16. The *De Lateribus* section of *De revolutionibus*. (See also the Copernicus facsimile published by Macmillan and the Polish Scientific Publishers, London–Warsaw–Krakow, 1972, of the manuscript: left-hand part of page 20.) Folio 19v of 1543 edition. (Photograph by Owen Gingerich. Permission of Harvard College Library.)

NICOLAI COPERNICI

De lateribus & angulis triangulorum planorum rectilineorum. Cap. XIII.

I.

Trianguli datorum angulorum dantur latera. Sit inquam, triangulum ABC, cui per quintum problema quarti Euclidis circumscribatur circulus. Erunt igitur & AB, BC, CA circumferentiæ datæ, eo modo, quo CCCLX partes sunt duobus rectis æquales. Datis autem circumferentijs dantur etiam latera trianguli inscripti circulo tanquam subtensæ, per expositum Canonem, in partibus, quibus dimetiens assumpta est 200000.

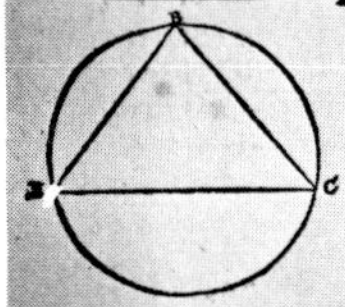

II.

Si uero cum aliquo angulorum duo trianguli latera fuerint data, & reliquum latus cū reliquis angulis cognoscetur. Aut enim latera data æqualia sunt, aut inęqualia. Sed angulus datus aut rectus est, aut acutus, uel obtusus. Ac rursus latera data datū angulum uel cōpræhendunt, uel non compræhendunt. Sint ergo primum in triangulo ABC duo latera, AB & AC, data æqualia, quæ angulum A datum compræhendunt. Cæteri igitur, qui ad basim BC cum sint æquales, etiam dantur, uti dimidia residui ipsius A, è duobus rectis. Et si qui circa basim angulus primitus fuerit datus, datur mox ipsi cōpar, atque ex his duorum rectorum reliquus. Sed datorum angulorum trianguli dantur latera, datur & ipsa BC basis, ex Canone in partibus quibus AB uel AC tanquam ex centro fuerit 100000. partium siue dimetiens 200000. partium.

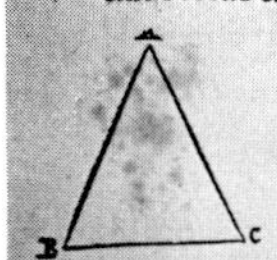

III.

Quòd si angulus, qui sub BAC rectus fuerit datis compræhensus lateribus, idem eueniet. Quoniam liquidissimū est, quòd quæ ex AB & AC fiunt quadrata, æqualia sunt ei,

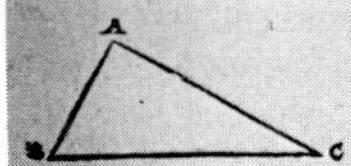

NICOLAI COPERNICI

SIGNORVM STELLARVMQVE DESCRIPTIO CANONICA, ET PRIMO quæ sunt Septentrionalis plagæ.

Formæ stellarum VRSAE MINORIS SIVE CYNOSVRAE.	Lōgitudinis partes.		Latitudinis partes	magnitudo
In extremo caudæ.	53 1/2		66 0	3
Sequens in cauda.	55 1/2 1/3		70 0	4
In eductione caudæ.	69 1/3		74 0	4
In latere q̄drāguli p̄cedēte australior	83 0		75 1/3	4
Eiusdem lateris Borea.	87 0		77 1/2 1/6	4
Earū quæ in latere sequēte australior	100 1/2		72 1/2 1/6	2
Eiusdem lateris Borea.	109 1/2		74 1/2 1/3	2
Stellæ 7. quarum secudæ magnitudinis 2. tertiæ 1. quartæ 4.				
Et q̄ circa Cynosurā informis in latere sequēte ad rectā lineā maxīe aust.	103 1/3		71 1/6	4

VRSÆ MAIORIS QVAM ELICEN VOCANT.

Formæ stellarum	Lōgitudinis partes.		Latitudinis partes	magnitudo
Quæ in rostro.	78 1/2 1/6		39 1/2 1/3	4
In binis oculis præcedens.	79 1/6		43 0	5
Sequens hanc.	79 1/2 1/6		43 0	5
In fronte duarum præcedens.	79 1/2		47 1/6	5
Sequens in fronte.	81 0		47 0	5
Quæ in dextra auricula præcedente.	81 1/2		50 1/2	5
Duarum in collo antecedens.	85 1/2 1/3		43 1/2 1/3	4
Sequens.	92 1/2 1/3		44 1/3	4
In pectore duarum Borea.	94 1/3		44 0	4
Australior.	93 1/3		42 0	4
In genu sinistro anteriori.	89 0		35 0	3
Duarū in pede sinistro priori borea.	89 1/2 1/3		29 0	3
Quæ magis ad Austrum.	88 1/2 1/6		28 1/2	3
In genu dextro priori.	89 0		36 0	4
Quæ sub ipso genu.	101 1/6		33 1/2	4
Quæ in humero.	104 0		49 0	2
Quæ in ilibus.	105 1/2		44 1/2	2
Quæ in eductione caudæ.	116 1/2		51 0	3
In sinistro crure posteriore.	117 1/3		46 1/2	2
Duarū p̄cedēs in pede sinistro poster.	106 0		29 1/2	3
Sequens hanc.	107 1/2		28 1/4	3

Quæ

FIG. 17. Star Table in *De revolutionibus*. The printed copy of the manuscript-page 52 of the facsimile (London–Warsaw–Krakow, 1972); folio 46v of 1543 edition. (Photograph by Owen Gingerich. Permission of Harvard College Library.)

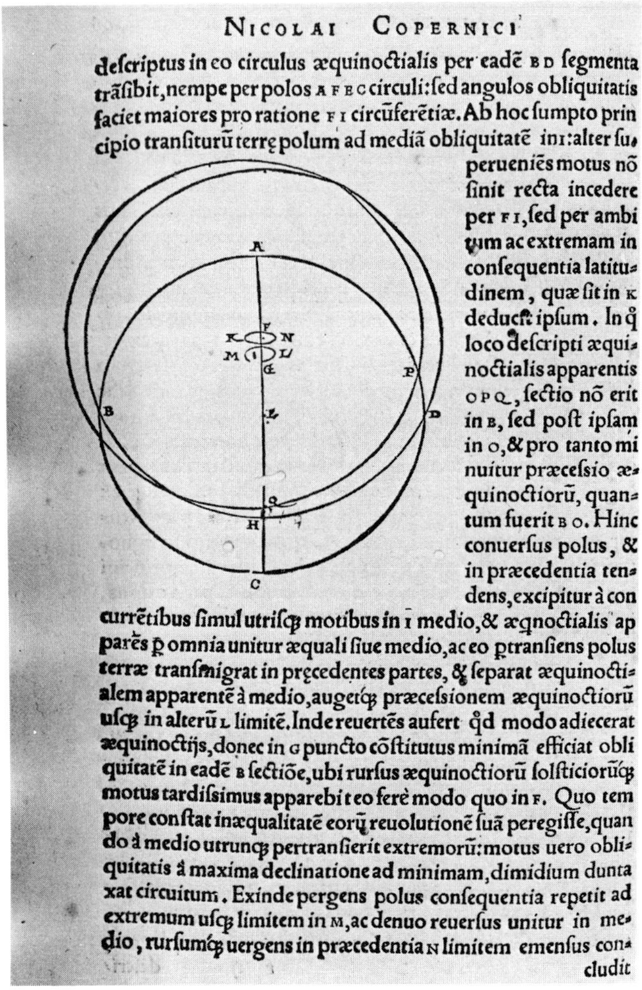

NICOLAI COPERNICI

descriptus in eo circulus æquinoctialis per eadē B D segmenta trāsibit, nempe per polos A F E C circuli: sed angulos obliquitatis faciet maiores pro ratione F I circūferētiæ. Ab hoc sumpto principio transiturū terrę polum ad mediā obliquitatē in I: alter superueniēs motus nō sinit recta incedere per F I, sed per ambitum ac extremam in consequentia latitudinem, quæ sit in K deducit ipsum. In q̊ loco descripti æquinoctialis apparentis O P Q, sectio nō erit in B, sed post ipsam in O, & pro tanto minuitur præcessio æquinoctiorū, quantum fuerit B O. Hinc conuersus polus, & in præcedentia tendens, excipitur à concurrētibus simul utrisq; motibus in I medio, & æqnoctialis apparēs p̄ omnia unitur æquali siue medio, ac eo ptransiens polus terræ transmigrat in pręcedentes partes, & separat æquinoctialem apparentē à medio, augetq; præcessionem æquinoctiorū usq; in alterū L limitē. Inde reuertēs aufert q̊d modo adiecerat æquinoctijs, donec in G puncto cōstitutus minimā efficiat obliquitatē in eadē B sectiōe, ubi rursus æquinoctiorū solsticiorūq; motus tardissimus apparebit eo ferè modo quo in F. Quo tempore constat inæqualitatē eorū reuolutionē suā peregisse, quando à medio utrunq; pertransierit extremorū: motus uero obliquitatis à maxima declinatione ad minimam, dimidium duntaxat circuitum. Exinde pergens polus consequentia repetit ad extremum usq; limitem in M, ac denuo reuersus unitur in medio, rursumq; uergens in præcedentia N limitem emensus concludit

FIG. 18. *De revolutionibus*; the print of page 74 of the Copernicus facsimile edition; folio 66v of 1543 edition. (Photograph by Owen Gingerich. Permission of Harvard College Library.)

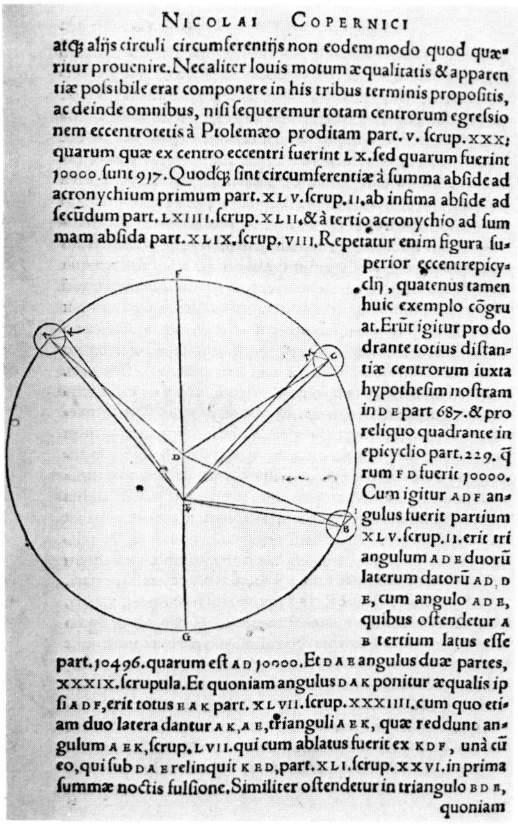

NICOLAI COPERNICI

atq; alijs circuli circumferentijs non eodem modo quod quæritur prouenire. Nec aliter Iouis motum æqualitatis & apparentiæ possibile erat componere in his tribus terminis propositis, ac deinde omnibus, nisi sequeremur totam centrorum egressionem eccentrotetis à Ptolemæo proditam part. V. scrup. XXX: quarum quæ ex centro eccentri fuerint LX. sed quarum fuerint 10000. sunt 917. Quodq; sint circumferentiæ à summa abside ad acronychium primum part. XLV. scrup. II, ab infima abside ad secūdum part. LXIIII. scrup. XLII. & à tertio acronychio ad summam absida part. XLIX. scrup. VIII. Repetatur enim figura superior eccentrepicycli, quatenus tamen huic exemplo cōgruat. Erūt igitur pro dodrante totius distantiæ centrorum iuxta hypothesim nostram in D E part 687. & pro reliquo quadrante in epicyclio part. 229. q̄ rum F D fuerit 10000. Cum igitur A D F angulus fuerit partium XLV. scrup. II. erit triangulum A D E duorū laterum datorū A D, D E, cum angulo A D E, quibus ostendetur A E tertium latus esse part. 10496. quarum est A D 10000. Et D A E angulus duæ partes, XXXIX. scrupula. Et quoniam angulus D A K ponitur æqualis ipsi A D F, erit totus E A K part. XLVII. scrup. XXXIIII. cum quo etiam duo latera dantur A K, A E, trianguli A E K, quæ reddunt angulum A E K, scrup. LVII. qui cum ablatus fuerit ex K D F, unà cū eo, qui sub D A E relinquit K E D, part. XLI. scrup. XXVI. in prima summæ noctis fulsione. Similiter ostendetur in triangulo B D E, quoniam

FIG. 19. Epicyclets diagram, printed in *De revolutionibus*. This is the print of page 165 of the facsimile edition; folio 150v of 1543 edition. (Photograph by Owen Gingerich. Permission of Harvard College Library.)

FIG. 20. Planetary position tables in *De revolutionibus*; the printed manuscript-page 184 of the facsimile edition; folio 174v of 1543 edition. (Photograph by Owen Gingerich. Permission of Harvard College Library.)

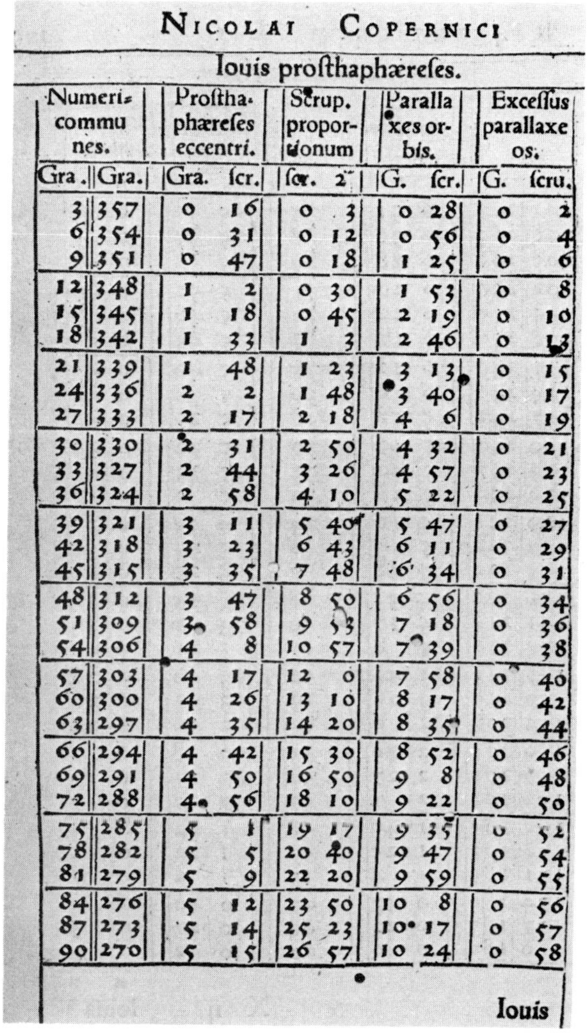

NICOLAI COPERNICI

Iouis prosthaphæreses.

Numeri communes.		Prostha- phæreses eccentri.		Scrup. proportionum		Parallaxes orbis.		Excessus parallaxeos.	
Gra.	Gra.	Gra.	scr.	scr.	2	G.	scr.	G.	scru.
3	357	0	16	0	3	0	28	0	2
6	354	0	31	0	12	0	56	0	4
9	351	0	47	0	18	1	25	0	6
12	348	1	2	0	30	1	53	0	8
15	345	1	18	0	45	2	19	0	10
18	342	1	33	1	3	2	46	0	13
21	339	1	48	1	23	3	13	0	15
24	336	2	2	1	48	3	40	0	17
27	333	2	17	2	18	4	6	0	19
30	330	2	31	2	50	4	32	0	21
33	327	2	44	3	26	4	57	0	23
36	324	2	58	4	10	5	22	0	25
39	321	3	11	5	40	5	47	0	27
42	318	3	23	6	43	6	11	0	29
45	315	3	35	7	48	6	34	0	31
48	312	3	47	8	50	6	56	0	34
51	309	3	58	9	53	7	18	0	36
54	306	4	8	10	57	7	39	0	38
57	303	4	17	12	0	7	58	0	40
60	300	4	26	13	10	8	17	0	42
63	297	4	35	14	20	8	35	0	44
66	294	4	42	15	30	8	52	0	46
69	291	4	50	16	50	9	8	0	48
72	288	4	56	18	10	9	22	0	50
75	285	5	1	19	17	9	35	0	52
78	282	5	5	20	40	9	47	0	54
81	279	5	9	22	20	9	59	0	55
84	276	5	12	23	50	10	8	0	56
87	273	5	14	25	23	10	17	0	57
90	270	5	15	26	57	10	24	0	58

Iouis

FIG. 21. Title-page of Erasmus Reinhold *Ephemerides*. Published in 1550. (Photograph by Charles Eames. Courtesy of the Uppsala University Observatory Library.)

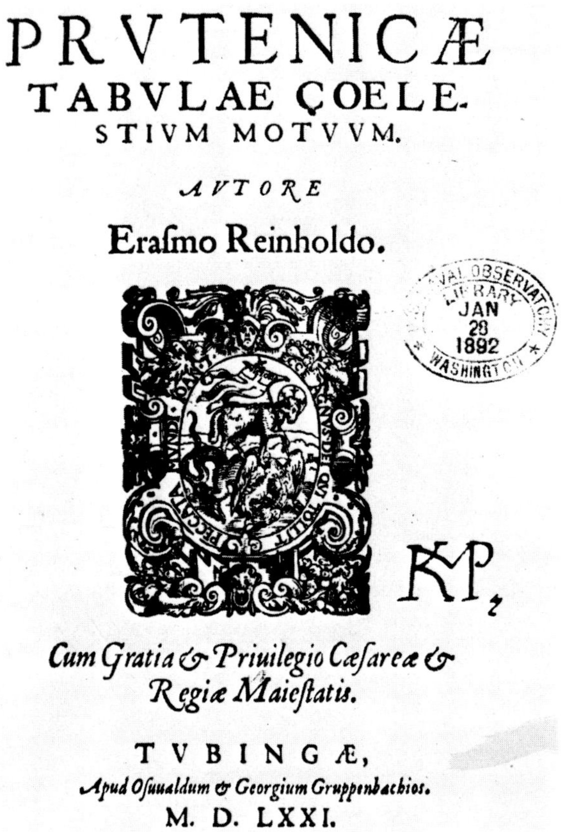

PRVTENICÆ
TABVLAE ÇOELE-
STIVM MOTVVM.

AVTORE
Erasmo Reinholdo.

NAVAL OBSERVATORY LIBRARY JAN 28 1892 WASHINGTON

Cum Gratia & Priuilegio Cæsareæ & Regiæ Maiestatis.

TVBINGÆ,
Apud Osuualdum & Georgium Gruppenbachios.
M. D. LXXI.

FIG. 22. Title-page of *Prutenic Tables* by Erasmus Reinhold. 1571 edition. (U.S. Naval Observatory Library.)

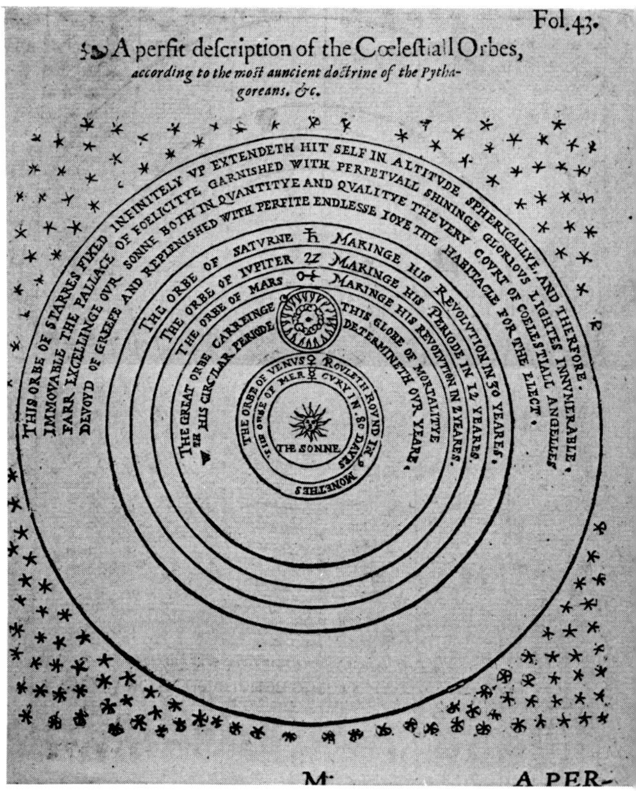

FIG. 23. Copernican Cosmology from Thomas Digges, *A Perfit Description of the Coelestiall Orbes*, published in 1576. (Photograph by Owen Gingerich. Courtesy of Harrison Horblit.)

FIG. 24. Tycho Brahe Cosmology drawing showing circular orbits of Mercury (LKMN), Venus (OPQR) and Comet of 1577 (STVX). In *De Mundi Aetherei Recentioribus Phaenomenis. Liber Secundis.* Printed at Uraniborg in 1588. (Photograph by Owen Gingerich. By courtesy of Harvard College Library.)

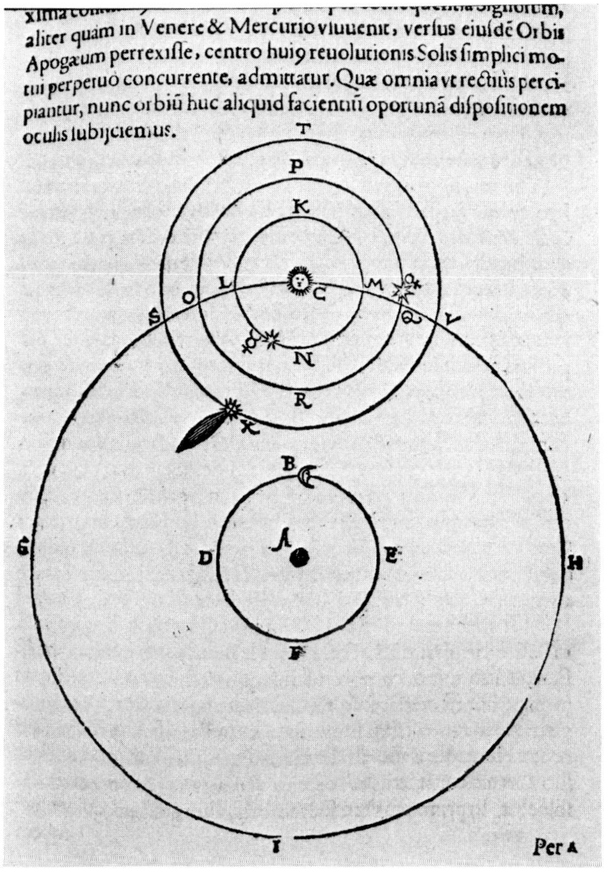

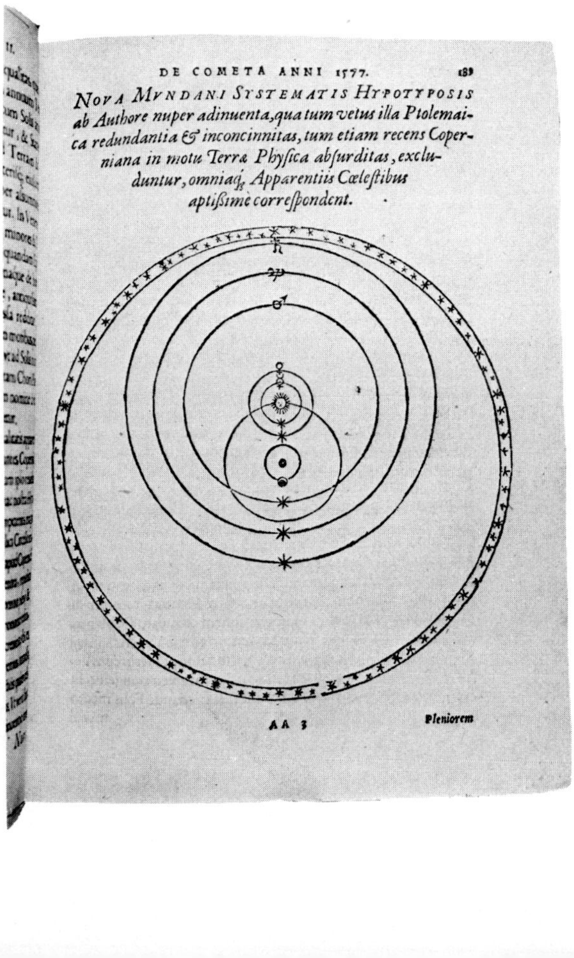

FIG. 25. Tycho Brahe Cosmology drawing in *De Mundi Aetherei Recentioribus Phaenomenis. Liber Secundis.* Printed at Uraniborg in 1588. (Photograph by Owen Gingerich. By courtesy of Harvard College Library.)

SESSION 1

The Scientific Significance of Copernicus

The Medieval Background to Copernicus

John North

University of Oxford

According to Aristotle, happiness was not to be found in the search for truth, but in the contemplation of truth previously attained.[1] An all too common view of the astronomer of the late Middle Ages is that he was complacent in his search for this sort of happiness—content, in other words, merely to retail the methods and truths of the ancients. Some writers will go even beyond this, and claim that as late as the time of Copernicus a proficiency in the mathematical methods of Ptolemaic planetary astronomy was something alien to European astronomy.[2] But then, at the other extreme, there are those who regard the medieval astronomer as having lived in a state of intellectual anxiety, constantly worrying about the imperfections of the Ptolemaic system.[3] The truth of the matter lies elsewhere, but it is neither pure nor simple, and in the frame of a short lecture I can scarcely begin to do justice to it. I can say nothing

[1] 1177[a] 26.

[2] See, for example, T. S. Kuhn, *The Copernican Revolution: Planetary Astronomy in the Development of Western Thought*, 2nd ed., 1969, pp. 123–4: "In fact there was almost no [developed planetary astronomy] in Europe during the Middle Ages, partly because of the intrinsic difficulty of the mathematical texts and partly because the problem of the planets seemed so esoteric. . . . Until two decades before Copernicus' birth in 1473 there was little concrete evidence of technically proficient planetary astronomy. Then it appeared in works like those of the German Georg Peuerbach (1423–1461) and his pupil Johannes Müller (1436–1476)."

[3] The imperfections, as Robert Small wrote, were however, "not immediately perceived, especially during the confusions which attended the decline and destruction of the Roman Empire"! *An Account of the Astronomical Discoveries of Kepler*, 1804, p. 81.

of the transmission of Hellenistic astronomy immediately after the time of Ptolemy (say A.D. 141), first to Persia and India, later to Islam, and finally to Europe, which it reached especially through contacts in Sicily and Spain. I shall speak mainly of the *Weltanschauung* of the Christian Europe to which Copernicus belonged. I shall make no attempt to break down his compound thoughts into atoms; much less shall I try to prove that each of the atoms was a constituent of the medieval cosmos, and that therefore Copernicus need astonish you no longer. I am simply here to set the scene, and, therefore, I shall speak of his intellectual precursors rather more than I shall speak of his intellectual debts. It is now fashionable to say that Copernicus gave us old wine in new bottles. However true, or untrue, this may have been of Copernicus, it is an apt description of my own contribution. I have nothing to say that is new, and the most I can hope is that my audience has not heard it all before.

Almost all of the astronomy known to the Christian scholar of the early Middle Ages came from one of seven authors, whose work, judged by the standards of Hellenistic astronomy, was extremely dilute. There were the encyclopedic works of Pliny (first century A.D.), Martianus Capella (fifth century), and Isidore of Seville (seventh century); there was a translation and commentary by Chalcidius (fourth century) of Plato's *Timaeus*; there was another commentary (fifth century) by Macrobius on a work called *Scipio's Dream*, by Cicero; there were the very influential writings of Boethius (sixth century); and there were the many works of Bede (eighth century). Even these seven writers were far from being mutually independent, but their apparently independent repetitions of one and the same idea served to reinforce confidence in that idea, however preposterous it might have been. Worse still, there was in all these writers a sorry absence of underlying explanation.

A more potent influence on the cosmology of early Christian Europe was that extensive body of writings left by the early Fathers of the Church. Their astronomical knowledge, even in aggregate, was pitifully small, and the strong desire they had to reconcile their cosmology with the Scriptures meant that even those of them who were familiar with Greek sources—men like Clement of Alexandria, Origen, Basil, and Augustine—were incapable of transmitting even the spirit of inquiry which had guided Eudoxos, Hipparchos, and Ptolemy. When we consider the allegorical tradition whereby the Earth was symbolized by the Tabernacle built by Moses in the wilderness, whereby the inner Tabernacle was made to correspond to the Kingdom of Heaven, and so forth, and when we consider the harassment of those who taught the sphericity of the Earth and the existence of the antipodes, then we can only marvel at the moderately enlightened attitudes of such Christians as Isidore and Bede, however trifling we ourselves happen to find their writings.

I have no wish to give the impression that there were all-pervading and official Church dogmas restricting cosmological thought throughout the Middle Ages, but we should never lose sight of the Church's monopoly of academic instruction, and of the ways in which all intellectual attitudes were to some extent coloured with religious belief. Man was, above all, a fallen creature, whose instincts were base, and whose reason was often at variance with those things in which he put his faith. With the Last Judgement, thought Augustine, the world would be newly organized, so why trouble to rationalize its present organization? For Isidore, the Fall led to a diminution in the intensity of the light from the Sun and Moon, but Isaiah assured us that after the redemption all would return to its former glory. Such instances of intellectual pessimism do not betoken a complete abandonment of the Greek idea that nature is orderly, intelligible, and accessible to human reasons, but they may be taken, for the purposes of my all too hasty lecture, to characterize the period before the eleventh century. Then, generally speaking, Christian man saw himself as a degraded and miserable creature whose only hope was in prayer and penitence, and as one to whom a rational understanding of the motions of the planets was supremely irrelevant.

This gloomy state of affairs brightened very rapidly with improvements in the organization of cathedral and monastic education. Secular schools were founded, which in turn led to the establishment of the first modern universities. At Bologna, Paris, and Oxford, religion was harmonized with other learning, especially that new learning which, from the twelfth century on, was flooding into Europe in the form of translations from Greek and Arabic. And then Aristotle arrived on the scene, not in a diluted Boethian form, but in respectable renderings of his own writings. Before long, Aristotle's power to stimulate thought became evident, and perhaps the greatest achievement of his greatest commentator, Thomas Aquinas, was to convince men that the universe, like God, could be understood through the operation of reason, and that the best of ancient learning was reconcilable with the Scriptures. It is no accident that in the most majestic of all medieval allegories, the *Divine Comedy*, which was written about a quarter of a century after the death of Aquinas, Dante should have provided a moral theme[4] within a framework of Aristotelian cosmology. Dante's *Paradise* is more than an unbroken succession of descriptions of heavenly bliss. It has an astronomical structure, and when the souls, for instance, are placed rank upon rank, the ranks are simply the planetary spheres.[5] In the *Convivio*, Dante refers to Aristotle as "that glorious philosopher to whom Nature most fully revealed her secrets",[6] and the poet goes on to make a remark which Copernicans will note with interest. Dante will not, he says, give the reasons which Aristotle alleges contradict those who claimed that the Earth is not fixed in its place. "It is enough for those whom I am addressing to know, on his weighty authority, that this Earth is fixed

[4] According to Dante (*Epistolae*, xi. 8), "the subject of the work as a whole, literally accepted, is the state of souls after death. . . . But if the work be taken allegorically, the subject is MAN, who makes himself liable, by the good or bad exercise of his free will, to the rewards or punishments of justice."

[5] *Par.* iii. 52. I am also convinced that Purgatory had a structure which was meant to reflect that of the heavens.

[6] *Conv.* III. v. 55.

and does not revolve, and that with the sea it is the centre of heaven." This passage should remind us how all those scholars who made commentaries on the *De Caelo* were clearly obliged to come to terms with possible motions of the Earth.

Dante's words also serve to illustrate a typical attitude to Aristotelian authority. On the other hand, the force of Aristotle's arguments was not such that Dante accepted them blindly, and he does contrive to disagree with Aristotle on at least two occasions, first as to the number of spheres, and second as to the number of Intelligences causing the circular movements of the spheres.[7] In the Renaissance, medieval scholastics were often accused of a slavish adherence to Aristotle; but coming from an age obsessed with the imitation of literary ornament no less than with philosophical argument, the challenge need not be taken too seriously. And there is no doubt about Aristotle's influence on the one man who is often said to have banished Aristotle from cosmological speculation—namely Copernicus.

Aristotelian cosmology has two essential ingredients, one a theory of natural motion, the other an explanation of the movements of the planets, an explanation developed from that first given by Eudoxos. This quite extraordinary planetary theory was so debased by the time of the Middle Ages that its explanatory power was almost totally lost. In fact as a purely predictive theory it was no longer needed, the Ptolemaic scheme having superseded it. All that remained was the crude metaphysical idea that the universe was divided into two concentric regions, each with a spherically symmetric structure. The Earth was at the centre, in the elementary region, which contained water above the element earth, air above the water, and fire above the air. The fire reached to the sphere of the Moon, where began the ethereal region, the heavens. Between the thirteenth and seventeenth centuries, by far the best known treatise of astronomy was *the Sphere* by Sacrobosco, and we may quote him on the nature of the heavens:

> Around the elementary region there is the ethereal, which is lucid and immune from all variation in its unchanging essence, and which turns in a circular sense with a continuous motion. It is called the "fifth essence" by philosophers. Of this there are nine spheres . . . namely of the Moon, Mercury, Venus, the Sun, Mars, Jupiter, Saturn, the fixed stars, and the final heaven.[8] Each of these spheres encloses the one below spherically.[9]

Sacrobosco's treatise was an elementary work on the astronomy of very simple phenomena, rising and setting, the zodiac, the spherical Earth, the climates, and the Sun's path. (It contained next to nothing on Ptolemy's planetary theory.) And yet

[7] See *Conv.* II. v. 20. Aristotle argued that if one Intelligence took care of each of the circular movements, that was enough. Any other intelligence would have been eternally useless, without any function. For Dante the mere *existence* of an Intelligence is its function.

[8] By a simple extension of the idea of concentric spheres, the region beyond the eighth, the empyrean, was itself often given a complex structure. This was done in several ways, that of Boethius being one of the best known. See Gregor Maurach, *Coelum Empyreum*, Steiner, Wiesbaden, 1968.

[9] Translated from Lynn Thorndike, *The Sphere of Sacrobosco and its Commentators*, 1949, p. 79.

even then, I strongly suspect that the passage I have quoted was as much as most educated men tended to remember of it, shall we say a year or two after they had left their university.

The second central doctrine of Aristotelian cosmology concerned natural motion. The quintessential, or fifth, element, out of which the celestial regions were made, was said to be not homogenous, being least pure where it borders on the air around the sublunary region. Its natural motion, circular and continuous, was to be contrasted with the rectilinear motions of the other elements; for earthly bodies fall naturally to the centre of the Earth, while fire rises naturally in the opposite direction, again radially outwards from the centre of the Earth, the centre of the Universe.[10] Forced motions were another matter, which we need not consider here.

Copernicus summarized this Aristotelian cosmology before giving his reasons, in the first book of the *De revolutionibus*, for thinking it inadequate.[11] He, after all, wished to place the Earth in the celestial category. On the other hand, he was enough of an Aristotelian to preserve the doctrine that each body has a natural place in the universe. His categories were Aristotle's reapplied: the Earth had a constant circular motion of planetary type and the same was true of its constituent parts, although a detached portion of Earth moved now with a compound motion,[12] having one naturally circular component and one naturally linear component directed to the Earth's centre. And now that the Earth was displaced from the centre of the Universe, Copernicus felt at liberty to do Aristotle the honour of generalizing his ideas:

> For my own part, I think that gravity is nothing but a certain natural striving implanted in the parts by the divine providence of the maker of all things, so that they may come together in the form of a globe, in unity and wholeness. It is believable that this striving is present also in the Sun, the Moon, and the other bright planets, so that they keep the resultant round shape which they show, even though they make their circular movements in many different ways.[13]

Those who suppose Copernicus to have cut astronomy free from the bonds of the Aristotelian doctrine that the universe is separated into two essentially distinct regions, in one of which matter circulates, and in the other of which it gravitates, must at least reconcile themselves to this passage.

Another occasion on which Copernicus found the spirit of the Peripatetic too strong to resist was in his rejection of the Ptolemaic principle of the equant. Whereas according to Ptolemy, most of the planets move uniformly around deferent circles, the uniformity of the latter motion is not with respect to arc length; that is to say, the

[10] Water and air were intermediate elements, whose heaviness and lightness, respectively, were less than those of earth and fire.

[11] See *De rev.* I. 7–8.

[12] Even in the idea of compounding motions he was following Aristotle and the medieval Aristotelians, and he is surely quoting Aristotle with approval when he avers that the division of motion into three sorts is merely an act of reason (*rationis solummodo actus*). *De rev.* I. 8.

[13] *De rev.* I. 9.

radius vector does not move round at a uniform rate. Instead, what rotates uniformly is the vector to the epicycle centre from a point some distance from the centre of the deferent circle—and both are, in general, distinct from the centre of the Earth.[14] In the eyes of Copernicus, the movements along Ptolemy's deferent circles were not uniform (*oportet inaequalem esse*).[15] I will not insist here that his mathematical imagination fell short of Ptolemy's, but merely note that he accepted as axiomatic what he called the "rule of absolute motion", according to which "everything would move uniformly about its proper centre",[16] and that this rule was in perfect conformity with the views put forward by Aristotle.[17]

Cosmology, in the Middle Ages, did not end with Aristotle and the extension of his scheme to an angelic hierarchy, nor did it end with the twin themes of Sun and Salvation as found in Dante. Such writings as Dante's, to be sure, had a reactionary effect, to the extent that they created in the mind of their audience an image of an all too simple universe, an image far too vivid to be easily put aside. Fortunately for astronomy, however, there were men for whom physical and metaphysical speculations were not enough, even in combination with theological and mystical allegory. The marked tendency towards a secularization of learning to be seen in the half century or so before *De revolutionibus*, was accompanied by a number of sound exegeses of the astronomy of Ptolemy's *Almagest*, but it would be a mistake to imagine that there was anything in them which would have surprised the best astronomers of the Middle Ages. Putting sources like Sacrobosco's elementary textbook aside, there are at least five categories of evidence for a competent level of astronomical activity in the Middle Ages.

First, the *Almagest* was far from being unknown *in extenso*. It was twice translated into Latin, in one case directly from the Greek, by an anonymous Sicilian writer in the middle of the twelfth century, and in the other case by Gerard of Cremona, in 1175. Another direct translation from the Greek was done by George of Trebizond in 1451, and this was to be printed in 1528, thirteen years after the printing of Gerard of Cremona's version. Needless to say, there is no fundamental difference between the three translations. In addition to these full translations there were several good epitomes of the *Almagest*, in some cases themselves translations out of the Arabic, but some of them made, so far as one can see, in Western Europe.[18]

[14] In his lunar theory, Copernicus obtained a more or less equivalent result by placing the Moon on a second epicycle, the centre of which was carried on the first; see below.

[15] *De rev.* IV. 2.

[16] *Commentariolus*, introduction; translation in E. Rosen, *Three Copernican Treatises*, 2nd edn., 1959, pp. 57–58.

[17] *De Caelo*, II. 6.

[18] One twelfth- or thirteenth-century epitome which was widely known, under one of a variety of titles (*Almagestum abbreviatum*, *Almagestum parvum*, and so forth), was transcribed and annotated by so late a writer as Regiomontanus. The work is currently being studied by Dr R. P. Lorch.

Second, there were the translations of a number of works showing some originality, and first written in Arabic. Such were the works of al-Battānī and Thābit ibn Qurra (both of the ninth century), al-Zarqālī (eleventh century) and Jābir ibn Aflaḥ (twelfth century). The last two astronomers mentioned here (Arzachel and Geber, as they were known in Latin) lived and worked in the Iberian peninsula, from which place also came a number of works in an Aristotelian tradition critical of Ptolemy, and especially of the epicycle mechanism. The alternatives offered involved a hypothesis broadly reminiscent of that of Eudoxos, requiring nested concentric spheres with non-coincident poles. The best known, but astronomically unsuccessful, work in this critical vein was by al-Biṭrūjī (Alpetragius), who died in 1235, and whose principal work was well known in the translation made in 1217 by Michael Scot.

Third, there were the tables used in astronomical calculation. The individual tables fall into many different categories, and a full set of perhaps a hundred tables usually derives from several different sources, although three extremely important and reasonably distinct collections were especially pervasive in Western Europe, namely that of al-Khwārizmī, the Toledan tables, and the Alfonsine. These seem to have circulated in Europe generally from, respectively, the early twelfth, the late twelfth, and the end of the thirteenth centuries. Although they were usually accompanied by canons, that is, treatises explaining in detail their use, they speak for a rather more intense intellectual training than is to be had from undergraduate courses in certain present-day academic subjects.

The fourth respect in which, I suggest, we have evidence of intelligent astronomical activity in the centuries before Copernicus is in the ancillary mathematics. That part of mathematics which we now call spherical trigonometry, came in a well-developed state into the repertoire of European learning through two principal channels, first, in the mathematical chapters of such works as Ptolemy's and Geber's, and second, in the canons to astronomical tables, and especially in Arzachel's canons. Such sources were actively enlarged upon by a number of European writers, of whom Richard of Wallingford, an early fourteenth-century English abbot, was one notable example, and Regiomontanus, more than a century later, another.

My fifth, and last, category of evidence concerns the design and fabrication of instruments, which were used both in observation and as an aid to calculation. (See Figs. 26–31.) There was the ubiquitous astrolabe, of course, with its continuous history from the Hellenistic and Roman world. This instrument was not only often modified in superficial and scientifically trivial stylistic ways, but was redesigned by two eleventh-century Toledan writers, 'Alī b. Khalaf and Arzachel, whose astrolabes were of universal application, in the sense that one plate sufficed for all geographical latitudes.[19] The *saphea arzachelis*, as the universal plate became known in the West,

[19] In principle, the conventional instrument requires a different local-coordinate plate for each latitude at which it is to be used.

was to be found, for example, on Richard of Wallingford's albion, an instrument by means of which very many different sorts of astronomical calculation could be made. I mention this last instrument partly because I am familiar with it, having recently edited the treatise devoted to it by its designer, but also because its complexity mirrors that of the astronomical learning of the time. At first sight resembling a rather large astrolabe, it contained upward of sixty different scales, four of which, for instance, comprised a spiral of thirty-one turns, rather as on many modern logarithmic slide rules. It was an observing instrument, and also permitted the calculation of eclipses and of planetary positions. These things it did in ways very different from instruments before it. There were broadly similar devices before this (all known under the generic name *equatorium*), and two European authors whose originality should not pass unnoticed here are Campanus of Novara and Peter of Dacia. The equatorium tradition was never broken, but rather began to fade away towards the end of the seventeenth century. There were, needless to say, several Copernican equatoria.

I cannot do more than hint at the very considerable medieval activity in instrument making, which extended to larger instruments for fundamental observation, as well as to such domestic items as pocket cylinder dials, ring dials, and so on. The art of gnomonics, which Copernicus himself was to essay, was certainly not ignored by medieval writers. Certainly a simple ring dial is very far from being a "world system", such as Copernicus is represented as having overthrown,[20] but its design implies a frame of mind which is altogether congenial to changing the frame of the world.

Medieval speculation on the motion of the Earth does not seem to me to have been significant in itself or to have been a very significant influence on Renaissance thought, but since, for the time being, we have all agreed to act as satellites to Copernicus, I will try at least to put in perspective what speculation there was. I have no wish to belittle medieval physics, but as an alternative to Aristotle (rather than as a more rigorously developed version of Aristotle) it seems to me to have failed in the important sense that the basic theory was not made to yield conclusions with the intention of putting them to empirical test. Speaking generally, the medieval scholastic had a penchant for offering all the potentially acceptable alternative solutions he could devise for a whole range of imaginary problems, and finally choosing one of these alternatives on metaphysical grounds. Now if there was ever any real prospect of loosening the Earth in its fixed central position, this was in regard to a possible rotation of its axis. Ptolemy had seen that, as far as appearances of stellar movements were concerned, the Earth might as well rotate on its axis as

[20] Cf. Alfons Kauffeldt's *Nikolaus Kopernikus: der Umsturz des mittelalterlichen Weltbildes*, Berlin, 1954.

be stationary.[21] He argued, however, that such terrestrial movements as had been proposed by the Pythagoreans Philolaos and Hicetas (both fifth century B.C.), and by Aristarchus (third century B.C.),[22] was ruled out by physical considerations—objects thrown up in the air would tend to be left behind, for example. Whether it was from the frequent discussions of Aristotle's *De Caelo*, or from reading Ptolemy's *Almagest*, one does not know, but at all events Jean Buridan in the fourteenth century stated that many people of the time thought it probable that the Earth moves on its axis with a daily rotation, the sphere of the stars being supposed stationary.[23] Like Ptolemy, he held that observed phenomena would be the same, and like Ptolemy he came down—and for the same reason—on the side of a fixed Earth; but he saw merit in the hypothesis of terrestrial rotation, for it seems the simpler of the two, the Earth being smaller than the sphere of the stars.[24] It has been suggested that Copernicus might have encountered Buridan's writings at Krakow, as a student, when Buridan's physical writings were required reading.

A younger contemporary of Buridan, Nicole Oresme, translated the *De Caelo* into French,[25] and wrote a commentary on it. It is unlikely that Copernicus was directly familiar with the commentary, although it might well have been the subject of contemporaneous debate. Oresme shared a disability with perhaps most of the scholastics: he could usually see both sides of an argument without being able to decide which side deserved his support. Like so many before him, he saw that from a kinematic standpoint there was no inconsistency in the idea of the Earth's rotation. He argued convincingly against Aristotle that movement of the heavens does not require the Earth to be at rest.[26] Experience, he says furthermore, cannot decide the

[21] *Almagest*, I. 7. Aristotle (*De Caelo*, 296^b) had used the same argument, but gave priority to two arguments of a different sort. He held that if the Earth were to move at all, it would move with both sorts of motion, which we may call revolution and rotation, since this is true of everything that moves with a circular movement, except the first sphere (*loc. cit.*). As a consequence we should find (i) that the stars would rise and set in different places, as do the planets, and (ii) the shapes of the constellations would change. These things are not observed, and so the assumption is false. Aristotle clearly assumed what Ptolemy did not, namely that the stars were near enough for his conclusions to be tested by experience.

[22] Aristarchus, as all who are familiar with the history of Copernicus will know, also believed in the revolution of the Earth in orbit round the central Sun, while the Pythagoreans had said that the Earth moved round a central fire, which provided the motive power for the whole Universe. Ptolemy, however, was discussing only axial rotation.

[23] *Quaestiones super libris quattuor de caelo et mundo*, ed. E. A. Moody, Cambridge, Mass., 1942, p. 227.

[24] Buridan's presentation of the case is not wholly consistent with this theory of impetus, but this is not of concern here.

[25] Completed 1377. The work, *Le livre du ciel et du monde*, has been edited by A. D. Menut and A. J. Denomy, Madison, Wisconsin, 1968.

[26] He shows that no movement—celestial or otherwise—requires in and of itself either the immobility or the movement of another body (*op. cit.*, pp. 372–3). Nicholas of Cusa later (unpublished treatise of 1444) assigned a rotation to both the Earth (which turned once a day) and the heavens (which turned twice, in the opposite sense, in the same period). See J. L. E. Dreyer, *History of the Planetary Systems from Thales to Kepler*, Cambridge, 1905, pp. 285–7.

matter.[27] And yet, after a long disquisition, in which it appears that he must come down on the side of Herakleides, who held that the Earth turned on its axis, he concludes with these words:

> However, everyone maintains, and I think myself, that the heavens do move and not the Earth: for God established the world which shall not be moved, in spite of contrary reasons. . . . What I have said by way of diversion or intellectual exercise can in this manner serve as a valuable means of refuting and checking those who would like to impugn our faith by argument.

I have been obliged to omit Oresme's eight pages of detailed argument, but his point has no doubt been taken: the Christian's faith was not a thing to be overturned by mental gymnastics. In ontological terms we may say that the world of the Faith was more real than that of physics, while for many—including Buridan and later Osiander—the world of astronomy was the least real of all. In Buridan's words, "For astronomers, it is enough to assume a way of saving the phenomena, whether it is really so or not."[28] When, in the Middle Ages, the reality of the celestial spheres was discussed—and there is no reason to suppose that Copernicus ever questioned their reality[29]—the discussion came within the province of physics rather than astronomy.

Another question on which the astronomer, as such, was not obliged to form an opinion concerned the bounds of the Universe, and its scale. Some genuinely Ptolemaic ideas on the scale of the Universe—concerning not only the *relative* sizes, but the *actual* dimensions of the orbs, quoted in miles—became widely diffused in Europe through the writings of al-Farghānī (or Alfraganus as he was known in the Latin West).[30] It is frequently said that Copernicus had a truly integrated planetary system, wherein the mean distances of the planets were fixed by his hypothesis that all motions were centred on the Sun. The fact is that the Middle Ages, too, had a system. It was arrived at by an *a priori* argument directed towards a harmonious overall scheme, and this it had in common with the arguments of the Copernican system. Of the two, the older harmony was the more deliberately aimed at; but it was wrong. The distance to the fixed stars was held to be approximately 130 million miles—

[27] *Ibid.*, pp. 520–37.

[28] *Op. cit.*, p. 229.

[29] Here it is noteworthy that there is no reason to suppose that Copernicus had first-hand knowledge of the works of al-Biṭrūjī, even in translation. See E. Rosen, "Copernicus and al-Bitruji", *Centaurus*, vii, pp. 152–6 (1961). Rosen reinforces Birkenmajer's thesis that Copernicus used the *Epitome of Ptolemy's Syntaxis* by Regiomontanus (ed. Princeps, Venice, 1496) as the principal source of his information about Muslim astronomy in general and Biṭrūjī in particular.

[30] When the relative sizes of the deferent and epicycle are known for a given planet, so too is the ratio of the maximum to the minimum distance from the Earth attained by the planet in its orbit. The scale of "Ptolemaic" distances is fixed by supposing that the planets are so nested that the maximum distance of one is the minimum distance of the planet next above. The unit of distance was determined in terms of the lunar distance. Apart from the entirely arbitrary nature of the hypothesis that the Universe is exactly filled by the planetary spheres, there is a weakness in the arguments leading to the Ptolemaic *order* of the planets. Cf. *De rev.* I. 10.

large, but certainly not infinite. Copernicus, however, was himself conservative in regard to the potentially infinite universe, as in so many other respects:

> First it is to be remarked that the universe (*mundus*) is spherical (*globus*), whether because this is of all shapes the most perfect, being an integral whole in need of no joins, or whether because this is the most capacious of figures, and therefore well suited to contain everything there is; or even because the separate parts of the world—the Sun, Moon, stars—are seen to be of such a shape; or because everything in the universe tends to be limited by this form—as is apparent with drops of water and other liquid bodies when they decide their own limits. . . .[31]

This was a conventional enough expression of a conventional view.[32] The decision as between a finite and an infinite universe was yet again one for the natural philosopher, rather than for the astronomer, as Copernicus pointed out with some resignation,[33] although in the same place he registered some surprise that a void beyond a finite universe should by some people have been conceived capable of holding together, as it were, a star sphere turning round with a violent motion.[34]

One of the besetting intellectual sins of the scholastics, that of forcing knowledge into neat categories, might not always have produced very profound results, but it is useful for us to examine occasionally the trees of knowledge delineated in the process. As always, Aristotle is lurking in the shadows, even where Boethius is ostensibly the centre of diffusion of his ideas. The three sorts of science, according to Aristotle, were theoretical (dealing with knowledge of mathematics, physics, and theology), practical (dealing with conduct) and productive (making beautiful or useful things). Mathematics was to be subdivided into the quadrivium of arithmetic, music, geometry, and astronomy—with which we are here concerned. Physics, or natural philosophy, was thus both on a different branch and at a different level, while a subject like calendar reform, to which Roger Bacon applied such astronomical knowledge as he had, would have been yet further removed, along with such "practical" sciences as astrology.[35] Suitably separated thus, astronomy became a subject for university study by most arts men, whereas astrology was rarely studied explicitly as part of the university curriculum, although, one suspects, it was studied assiduously on the side. It was never quite respectable to say that the *purpose* of

[31] *De rev*. I. 1.

[32] It was not always clear, however, whether the spherical world was embedded in an infinite void or whether space somehow ended with the spheres. Cf. Oresme, ed. Menut and Denomy, pp. 368–9, and note Oresme's remark that to contradict the proposition that the world actually moves with right motion through an imaginary infinite and motionless space was to maintain an article condemned in Paris. (The proposition would have God's power be limited. Copernicus here left God out of his arguments.)

[33] *De rev*. I. 8.

[34] But this he says in the course of demolishing an earlier argument, and he gives no indication of his own views as to what is beyond the sphere of stars.

[35] On the requirements of the Church and the Lateran Council in relation to calendar reform, see *De rev*., dedication to Pope Paul III.

studying astronomy was that one might apply it to astrology, but there is no doubt that this attitude was commonplace, especially in the fourteenth century and after, and no less after Copernicus than before him. Copernican astrology was slow to emerge, perhaps because most astrologers were feeble practitioners of mathematical astronomy, or perhaps because the very essence of astrology was to be ancient and mysterious.

Copernicus was no astrologer, but he uses astrological imagery, perhaps unwittingly, when he adapts the common renaissance theme of an equivalence of the Sun, God, the king, and a throne. He quotes Hermes on the Sun as the visible God, and shows a reverence for that Pythagorean tradition which would have the Sun in the centre because it has perfection and is the source of our light and life. This would have been out of place in the Middle Ages. A medieval writer, following Aristotle, might well have referred to Pythagoras, or to his pupil Herakleides, as we have seen, but he would not have done so with the abandon that became so common after the revival of interest in Platonism in the later fifteenth century.[36] Copernicus followed no author's example so sincerely as he followed Ptolemy's,[37] but this was not high fashion in 1539, when Rheticus in the *Narratio Prima* averred that his teacher had followed "Plato and the Pythagoreans, the greatest mathematicians of that divine age".[38] In 1542, when Copernicus wrote the dedication to his greatest work, he quoted Pliny, because Pliny referred to Philolaos (in connection with the Earth's movement round a central fire), and to Herakleides and Ekphantos (in connection with the Earth's rotation). This is purely and simply adulation regardless of rational argument, and adulation of a kind not easily found in the supposedly authoritarian Middle Ages.

Having said all this, I have to admit that, in his Pythagoreanism, Copernicus was doing little more than indulge in a tiresome literary conceit of a significance easily overrated. It is not my brief to ask how solid was the achievement beneath the literary veneer, but merely to contrast the medieval approach to astronomy with his. It is easy to argue for continuity, and to maintain that the Copernican revolution is an illusion since precious few new observations were involved in it, and especially since the theoretical structure of the new scheme was of the same kind as that of the old. Ptolemy was known to the Middle Ages, in the sense that his principal works were available, but he was generally unheeded in one important particular, that is to say, in his techniques for applying to observational data a well-controlled hypothetico-deductive method, finally leading to a planetary model and its parameters. Admittedly

[36] Copernicus was an assistant to the Platonist Domenico Maria Novara (*ca.* 1500: see the *Narratio Prima* of Rheticus, trans. Rosen, p. 111) and pupil of a leading humanist, Albert Brudzewski.

[37] Ptolemy comes in for some qualified but genuine praise in *De rev.* I. Prol.

[38] Rosen, *op. cit.*, p. 147.

Copernicus' creative powers ranged less freely than Ptolemy's.[39] The second order epicycle was a far more conservative construct than was Ptolemy's equant;[40] and yet each was the result of thought processes which were almost unknown in medieval Europe, where epitomes of the *Almagest* tended to give only Ptolemy's conclusions. Medieval astronomy was a vigorous activity, but the vigour of its practitioners was directed at making it more comprehensible, more easily applied to practical ends, more concisely expressed. Truth, once established, was not to be tampered with. In the words of a question asked by Dante:

> For what fruit would he bear who should demonstrate once more some theorem of Euclid; who should strive to expound a new felicity, which Aristotle has already expounded; who should undertake again the apology of old age, which Cicero has pleaded? Naught at all, but rather would such wearisome superfluity provoke disgust.[41]

Copernicus managed to look back beyond the Middle Ages, but he could not shake himself completely free from its influence. Occasionally he made use of a consistent sexagesimal system, as opposed to a mixed system (such as we use when taking zodiacal signs of 30°). But one way of representing the Alfonsine tables had been similarly consistent long before. He occasionally changed to a decimal notation; but again there were several medieval precedents. Copernicus is often, and not unreasonably, given credit for his explanation of the slow apparent drift of the equinoxes as a consequence of the movement of the Earth's axis, rather than as a movement of the sphere of fixed stars. But he was not able to appreciate the erroneous character of the prevailing medieval belief in what was known as trepidation, an irregularity in the movement of the equinoxes which was accounted for by a complex of circular movements in three dimensions. Copernicus adopted a model for which he clearly borrowed from the traditional models usually associated with the names of Thābit and Arzachel. There are occasions when it seems to be impossible to decide whether Copernicus was aware of his precursors or not. His secondary epicycle for

[39] Copernicus, of course, had the inestimable advantage of having Ptolemy's example to follow. He deliberately kept as close as possible to the conventional Ptolemaic data for the solar distance, and—at least for much of the time—to the Ptolemaic ratio 13/5 for the ratio of the diameters of the Earth's shadow (at the maximum lunar distance) and the Moon. See the masterly study of Copernicus' dependence on Ptolemy by Otto Neugebauer, "On the planetary theory of Copernicus", *Vistas in Astronomy*, Ed. A. Beer, vol. 10, pp. 89–103 (1968), especially p. 101. Even in the theory of planetary latitudes, where there are distinct advantages in the heliocentric approach, Copernicus followed the Ptolemaic method, in which the deferent planes of the planets are taken to pass through the Earth, thus wrongly introducing into the reckoning of latitudes the two eccentricity vectors (of the Earth and of the planet). Cf. Neugebauer, *op. cit.*, p. 103.

[40] There is a sense in which Copernicus preserved the equant. Both models are virtually equivalent, geometrically speaking, and Copernicus clearly aimed at making the discrepancy as small as possible. The planetary longitudes predicted in the two systems should all differ by considerably less than a minute of arc. See Neugebauer, *op. cit.*, pp. 92–95. Kepler explicitly reintroduced the equant into a new heliocentric model, and in doing so earned important dividends, for he was thereby led first to an oval orbit for Mars, and eventually to his elliptical orbits.

[41] *De monarchia*, I. i; trans. by P. H. Wicksteed, pp. 127–8, London, 1904.

the lunar motion had been previously used by Ibn ash-Shāṭir of Damascus, a century before Copernicus was born.[42] Nāsir ad-Dīn aṭ-Ṭūṣī, who worked at Marāgha and who died almost exactly two centuries before the birth of Copernicus, devised a geometrical construction for obtaining a linear motion from two circular motions, a construction which was also to be adopted by Copernicus.[43] Even if a day comes when, at last, it proves possible to trace a connection between the two Muslim astronomers and Copernicus, his debt to the Middle Ages is never likely to seem greater than his debt to Ptolemy; the very structure of *De revolutionibus*, let alone its detail, reflects that of *Almagest*. This perhaps explains in part why for so many historians, astronomy was a sort of Rip Van Winkle, sleeping soundly from late antiquity, until it was roused in 1543. (It is perhaps only a matter of time before someone writes a story in which Copernicus was a reincarnation of Ptolemy.) This sort of history is all very well, as long as it is recognized for what it is, namely the pure propaganda of humanism. The plain truth is that astronomy in the Middle Ages, like the ordinary people of the Kaatskill Mountains, was for most of the time very much awake.

Editorial Note

Illustrations of relevant astronomical instruments (Figs. 26–40)—covering the period A.D. 1054 to 1598—are reproduced on pp. 17–25.

[42] See Victor Roberts, "The planetary theory of Ibn al-Shāṭir: latitudes of the planets", *Isis*, vol. lxvii, pp. 208–19 (1966), and earlier papers in the series there listed.

[43] See Neugebauer, *The Exact Sciences of Antiquity*, pp. 203–4, 207, Providence, 1957.

FIG. 26. Gear-work from the interior of an Islamic astrolabe of A.D. 1221–2. This instrument worked by the gearing depicts the relative positions of the Sun and Moon in the zodiac. The shape of the teeth of the gears is characteristic of those of the earliest mechanical clocks in the West. (Courtesy of the Museum of the History of Science, Oxford.)

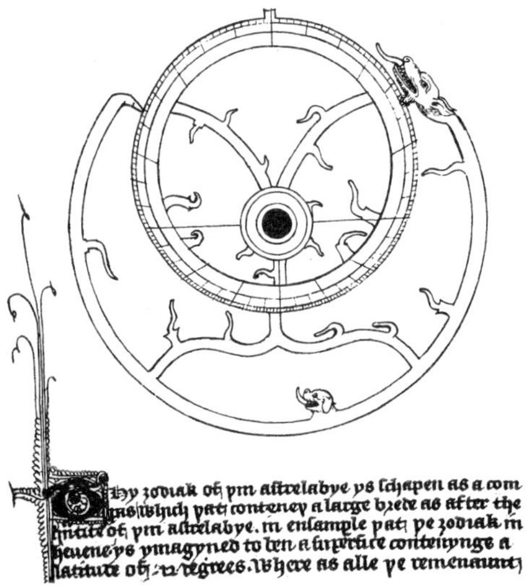
Thy zodiak of þin astrelabye ys schapen as a com
pas which þat conteneþ a large brede as after the
quantite of þin astrelabye. in ensample þat þe zodiak in
heuene ys ymagyned to ben a superfice conteynynge a
latitude of 12 degrees. Where as alle þe remenaunt

Fig. 27

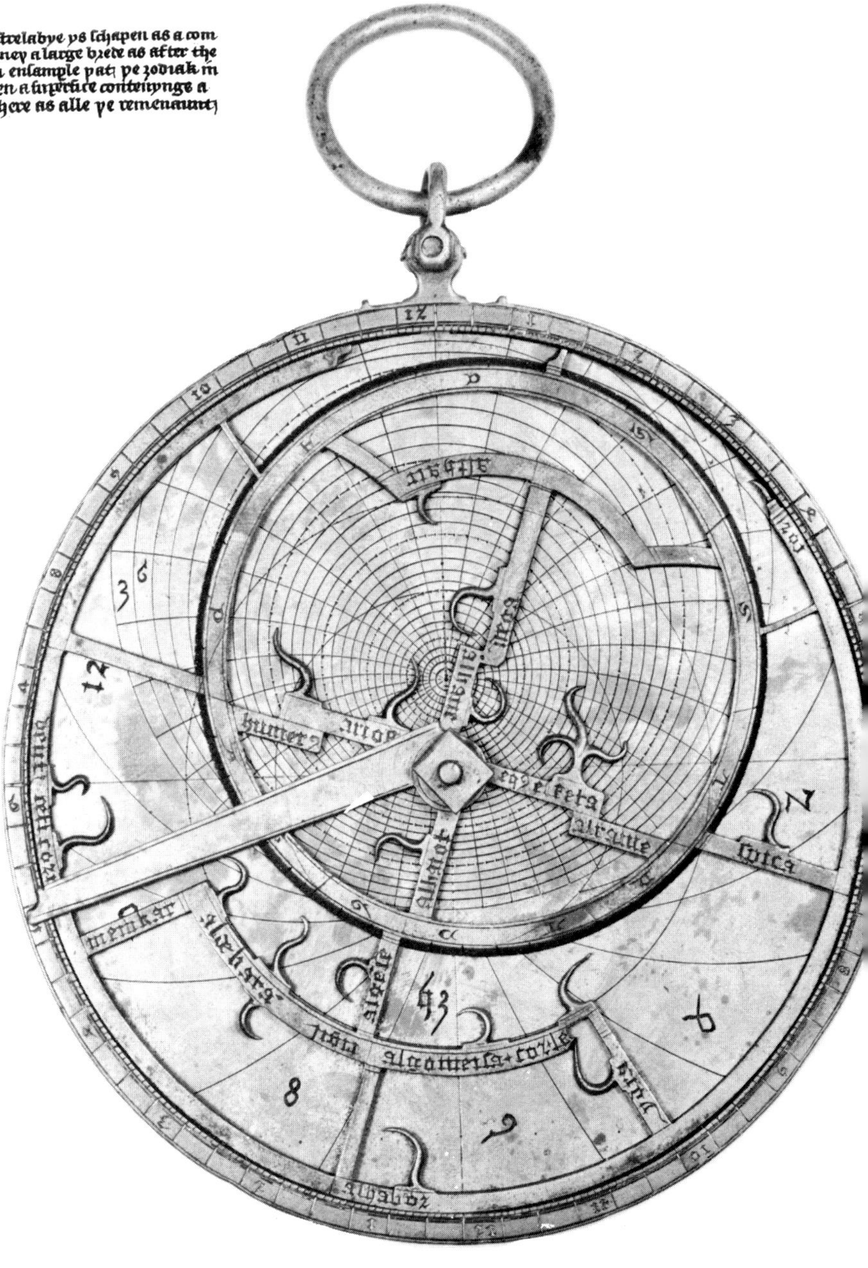

Fig. 28

Fig. 29

FIG. 27. The Rete of an astrolabe, as drawn in an early-fifteenth-century copy of Chaucer's treatise on the instrument. (Courtesy of the Bodleian Library, Oxford, M.S. Rawlinson D.913.)

FIG. 28. A fifteenth-century astrolabe of a sort characteristic of French workshops of the time. (Courtesy of the Museum of the History of Science, Oxford.)

FIG. 29. A fourteenth-century astrolabe Rete, unusual in that it is geared round its rim, conceivably for use on the dial of an early clock. (Courtesy of the Museum of the History of Science, Oxford.)

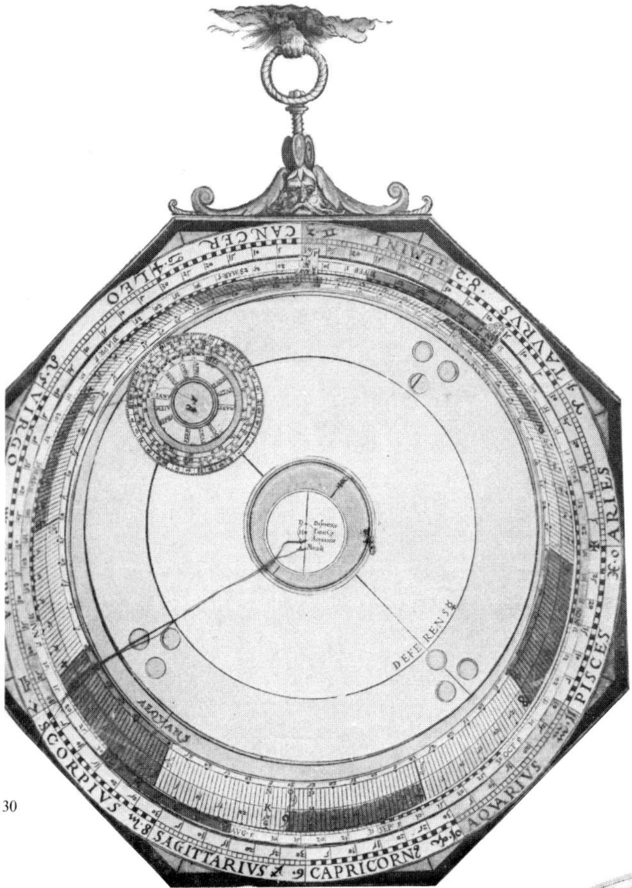
CANCER
LEO
GEMINI
TAVRVS
VIRGO
ARIES
PISCES
SCORPIVS
SAGITTARIVS
CAPRICORN
AQVARIVS
DEFERENS
AEQVANS

Fig. 30

Fig. 31

Fig. 32

FIG. 30. An equatorium of printed paper volvelles, from Peter Apian's *Astronomicum Caesareum*, Ingolstadt, 1540. The book contains many comparable devices. This example is for calculating the position of the planet Mercury. (Photograph by courtesy of Dr. John North, University of Oxford.)

FIG. 31. An equatorium of *ca*. 1600, which incorporates evidence of the use of *De revolutionibus*. The side shown is for calculating the positions of the Sun and the Moon, whereas the other side is for the five planets and is more closely related to the medieval equatorium working on Ptolemaic principles. (Courtesy of the City of Liverpool Museum.)

FIG. 32. Small Islamic astrolabe from A.D. 1054 in the Jagiellonian Museum in Krakow.

FIG. 33. Celestial Globe of Martin Bylica of Olkusz (1480) in the Jagiellonian Museum in Krakow.

FIG. 34. Detail from the Globe of Martin Byli Olkusz (1480), showing the constellation Au (Courtesy of The Warburg Institute, Universi London.)

Fig. 35. Martin Bylica's large European astrolabe from 1486 in the Jagiellonian Museum in Krakow.

Fig. 36. Martin Bylica's Torquetum from 1487 in the Jagiellonian Museum in Krakow.

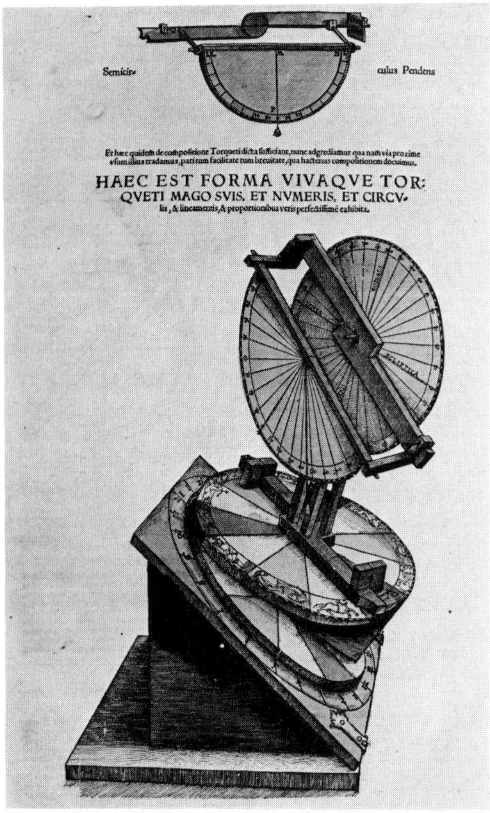

FIG. 37. The Torquetum from Peter Apian's *Astronomicum Caesareum*, Ingolstadt, 1540. (Photograph by Charles Eames.)

FIG. 38. Tycho Brahe's great brass globe described in *Astronomiae Instauratae Mechanica*. Date 1584, but originally constructed in Augsburg in 1570. The Globe is 149 cm in diameter. (Photograph by Charles Eames.)

FIG. 39. Quadrant described in *Astronomiae Instauratae Mechanica* by Tycho Brahe. Published in Wandesburg, 1598. The instrument was constructed in Augsburg in 1572. (Photograph by Charles Eames.)

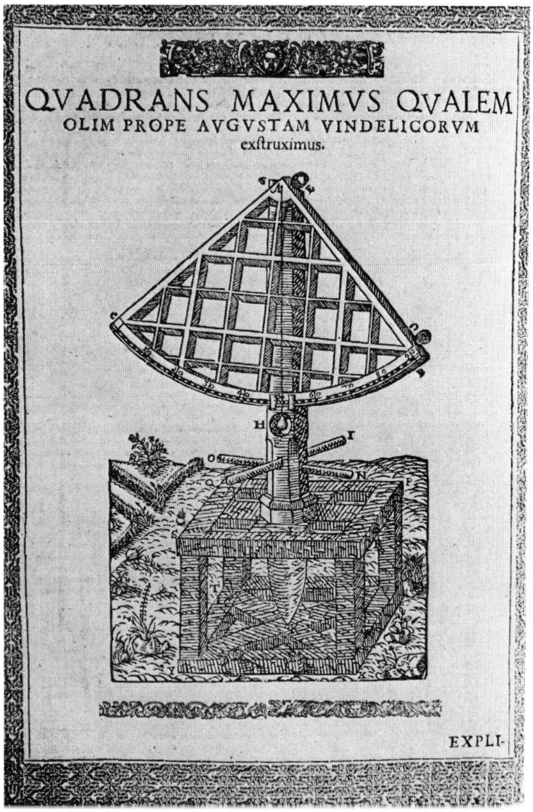

FIG. 40. Tycho Brahe's Equatorial Armilla described in *Astronomiae Instauratae Mechanica* (Wandesburg, 1598). (Photograph by Charles Eames.)

The Role of Observations in the Work of Copernicus

JERZY DOBRZYCKI

Institute of History of Science and Technology, Warsaw, Poland

FIVE hundred years ago an astronomical observation consisted of a direction to a celestial body, with a statement giving angular coordinates and time. An observation could also be a statement of a special and/or time coincidence of two different celestial objects. Thus an observation of a planet in reference to the star background establishes a relation between the theory of the motion of the planet (its orbital coordinates) and the equatorial (or ecliptical) coordinates on the sphere. An eclipse means that there is a coincidence of a set of coordinates of the Moon and the Sun. A time determination is usually included, but may not be necessary. At the time of Copernicus there were also observations in the form of a planet's position differing from its tabulated value by a stated number of degrees.

A rather delicate methodological point may be raised concerning the heuristic value of the observations. If an observation is made to prove a previous theory of a physical phenomena with which it is associated, then the outcome might be influenced by the assumptions which were involved. Fortunately this does not apply to all cases, as the history of science demonstrates. It is, however, the case when the observations are referred to the cosmological question of geo- versus helio-centric theory.

The observational equivalence of geostatic and geokinetic astronomy was well known to earlier scholars such as Nicole Oresme (1377) as well as to Copernicus himself.

Repeatedly, he remarks in *De revolutionibus* that the motions of the observer and of the observed can produce the same phenomena. This, however, does not mean that he considered all motions to be relative, as is often claimed. The direct arguments such as the aberration of light, stellar parallax and the progress of cosmogony of the solar system were to come only in the following centuries.

Therefore, when Copernicus used ancient as well as his own observations in his work it was not, and could not be, for the purpose of directly disproving the ancient ideas on the structure of the Universe.

The preserved records testify to more than sixty observations made by Copernicus, with almost half of them quoted in *De revolutionibus*. As an observer, Copernicus can be considered to be "average" with the precision of his observations varying from case to case. Surely, the best observations were those made without instruments, such as the observations of the planets for which he used the method of alignment to the nearby stars. The method was still used in the early hectic days of tracking artificial earth satellites in the 1950s.

We find evidence of Copernicus' creative interest in the techniques of observations not only in his printed work. Thus the preserved remnants of a solar observing table in the Olsztyn Castle show Copernicus as the designer of a theoretically interesting instrument. A net of hyperbolic lines mark the path of the solar image, reflected on the wall by a horizontal mirror; the lines are drawn for every fifth day, for a period of approximately a month before and after the equinoxes. Modern views of the relevant Frombork buildings are shown in Figs. 41–43 on pp. 30–31.

There is also indirect evidence (in the writing of Erasmus Reinhold, of Wittenberg) of Copernicus using the camera obscura principle for observing solar eclipses.

Copernicus was far more conservative in regard to observations and observational technique in his published work of *De revolutionibus.*

The instruments described in this work—the parallactic triangle, the spherical astrolabe, and the solar quadrant—are the Ptolemaic ones, and the descriptions of them are paraphrased from the *Almagest*. It is again the theoretical side of observations which shows the intellectual strength of Copernicus in the selection of objects to be observed for the explicit purpose of checking and/or revising the main parameters.

The case of the motion of the apsidal lines is an outstanding example. A set of three determinations (and not necessarily of high precision) of the opposition of a planet was needed to redetermine the spatial orientation of its orbit. Since Ptolemy, no observing astronomer had made the effort to make these observations. It was left to Copernicus between 1514 and 1529 to plan and execute a series of opposition observations. The result was the discovery of the motion of the apsidal lines.

Observations of this kind were planned and carried out by Copernicus for the purpose of renewing the foundation of practical astronomy based on new, revised parameters obtained through comparison of his observations with those of ancient astronomers, especially those of Ptolemy.

As a reinvestigation of the final tables in *De revolutionibus* shows, this revision of numerical data was only partially successful in that the improvement in the predictions of astronomical positions was obtained for only some phenomena. The Copernican tables, and later tables based thereon, became obsolete with the increased observational accuracy at the turn of the sixteenth century. However, in the meantime, it was the viewpoint of the work of Copernicus, by the restoration of the methods and the parameters of practical astronomy, which earned him fame as the most outstanding astronomer, "the second Ptolemy". In fact the practical, numerical side of *De revolutionibus* was eagerly accepted by scholars who could not accept the more important cosmological contents of the book, and the Copernican mathematical astronomy was revised at least twice to the geocentric form (e.g. by Giovanni Antonio Magini).

Another function of the observations in the work of Copernicus was for instructional purposes. Following Ptolemy, in giving an exposition of astronomy in general, Copernicus described not only the geometrical reasoning, with demonstrations and computations, but he also presented the method of obtaining the observational data. This purpose was served at times by "desk" observations such as the imaginary examples of lunar eclipses in Book IV, or by actual executed observations for determining the lunar parallax—which clearly show the geometry involved but, for practical reasons, are unsuited for obtaining reliable results.

Also, the opposition observations previously cited may be mentioned here, as they are presented in *De revolutionibus* with an idealized precision, obviously to fulfil the didactic purpose.

The main, most important role of the observations is their connection with the fundamental ideas of the Copernican astronomy—the new cosmology. In almost every section of the book the reader is given evidence that no observed facts could contradict the new theory and that they were all compatible with the fundamental concept of the earth moving in its daily, annual, and precessional motion.

Together with other "positive" aspects, like the discovery of order and harmony in the planetary system, this indirect evidence of observations formed a new, solid frame of reference for building up the "modern astronomy". This work was done by a sequence of generations of scientists, to whom Copernicus addressed *De revolutionibus*, asking them (freely quoted) "to consider thoroughly what is explained and shown in this, my book".

FIG. 41. Frombork (Frauenburg); Copernicus' observation tower.

FIG. 42. Walls and outer gate of Frombork Cathedral, where Copernicus was a canon. (Photograph by Charles Eames.)

FIG. 43. Frombork Cathedral, where Copernicus served as a canon while working out his heliocentric theory. (Photographed by Owen Gingerich in August 1972 during a meeting of the Copernicus Committee of the International Union for the History and Philosophy of Science.)

DISCUSSION

EDWARD ROSEN: It is not lightly that I undertake to differ with my good friend, Professor North, about one feature of his excellent presentation this morning. However, I should like to recall your attention to the passage in which Professor North discussed this very important question in the history of cosmology whether the Aristotelian dichotomy between a celestial world and a terrestrial world persisted in Copernicus' thought or was overthrown by him. I may not have caught the precise drift of Professor North's discussion of this matter, but if you will allow me I should like to say a few words on this subject, because I find that the discussion in print—I cannot read the text very carefully and be sure that I understand what the author is saying—is, it seems to me, contrary to the simple facts.

You will recall that Professor North reminded us of two essential features of the Aristotelian cosmos, to wit, that a circular motion is natural to the celestial bodies whereas rectilinear motion is natural to terrestrial objects. But since Copernicus moved the earth out of the terrestrial sphere, into the celestial sphere, strictly speaking there was no longer any terrestial sphere. It seems to me therefore inescapable that Copernicus must be recognized as the man who abolished the dichotomy between the terrestrial and celestial spheres. You may say, "Well, this is merely a logical inference that is made by students of Copernicus after his death". However, an attentive reading of his text shows that he was aware of what he was doing, but afraid of the practical consequences, for the reasons so eloquently set forth this morning by Professor North.

Let me just sketch briefly one little feature of the Copernican revision of Aristotelian physics. You will notice that Copernicus is not eager to depart from Aristotle just as he is not eager to depart from Ptolemy. Yet on account of his basic tenet, such departure is coerced, and this is the coercion. Let that arc represent a portion of the motion of the planet Earth, in the heavens, and let us propose that a detached bit of earth is moved in what was called violent or forced motion. Push your enemy off the roof—that is an excellent example. I do not recommend it to anybody by way of practical advice, but it provides us with a theoretical example. What will your enemy do as he falls from the top of the Washington Monument? He will partake of the circular motion of planet earth, of which he still is a part, and at the same time on account of the violent motion which you have imparted to him, he will display a rectilinear descent. We have here, therefore, a combination of two motions in a single projectile, which as you know of course becomes the hallmark of the beginning of modern physics, and we have simultaneously the combination of the two motions previously considered absolutely independent of each other. Thus circular motion in Aristotelian cosmology and physics belonged exclusively to a celestial body, while rectilinear motion belonged exclusively to a terrestrial object. But here in the case of your enemy being tossed from the roof of the Empire State building, you have the overthrow of the Aristotelian dichotomy. The same object descending from the roof of the Empire State building is simultaneously performing a circular revolution and a rectilinear descent. Therefore, the distinction between the two spheres—celestial and terrestrial—has been abolished. This seems to me to be crucial, if we are really to understand what Copernicus did. I may say one last word with regard to his careful avoidance of explicitness on this subject. The idea is clearly present but the formulation is not obvious. I think that the reason for the non-obviousness of the presentation should be evident to us today who reflect on what goes on in various countries in the world—what happens to people who disagree on all sorts of questions with those in power. If we reflect that the situation in Copernicus' day was much more acute than it is today, I think we can sympathize with his desire not to be too clear or explicit on this point.

OWEN GINGERICH: I am sure that this type of commentary could raise a very considerable debate here because of what may or may not be implied as a logical extension of what Copernicus put explicitly. I think that we see examples in other areas of history of science, for example, what Newton thought that Galileo should or may have done. I think Professor North has the right to make a remark or two in response.

JOHN NORTH: I will be very brief. I would disagree with absolutely nothing of what Professor Rosen has said. I hope I did not overstate my case, but the particular point I wanted to make was simply that Copernicus consciously generalized one of Aristotle's ideas. In a sense, a man who generalizes someone else's ideas and throws a bit away in the process is overthrowing him. For Aristotle, things below the sphere of the Moon gravitated towards the centre of the Earth. (Above the Moon, natural movement was different, being namely circular.) The result was a globular earth. In the passage I quoted from Copernicus, the idea is generalized, and it is said that in the Sun, the Moon, and the other bright planets there are perhaps similar gravitational forces. This ran counter to the Aristotelian doctrine as a whole, certainly, but in a sense it was a generalization of one Aristotelian idea of central importance. This very limited point was the only one I wished to make, and I fully share Professor Rosen's concern over the irresponsibility of those who would argue that Copernicus did nothing new.

NOEL SWERDLOW: In his manuscript, Copernicus frequently changes either the longitude or the time of an observation, correcting it from one value to another. I wonder if Professor Dobrzycki has investigated what these changes mean.

JERZY DOBRZYCKI: The cases I am thinking of are the oppositions of the planets. It would apply to Saturn where the time of an opposition is corrected once or twice in the manuscript and shows that the final epoch of the opposition has been recomputed using a more crude determination than was actually made. This was done so the epochs could be fitted into the tabulated parameters.

NOEL SWERDLOW: Copernicus is going from his theory back to his observations, making the observations agree with the computed positions.

OWEN GINGERICH: Let us consider the case of observing an opposition. How do you know operationally when the planet is in opposition? You must already know something about the solar theory in order to get this. This explains why in the work of Kepler it took him seventy trials to fit the circular orbits to the observations of Mars. When you go back to his calculations book, you find that he manages to carry out an iteration with an average of about fourteen trials. But in between the sets of trials he has to interpolate the original observations to a new assumed time of opposition. The oppositions finally tabulated are not necessarily actual observations. (I have written this for the International Kepler Symposium at Weil der Stadt.) And I think that could apply here in the case of Copernicus. Dr. Swerdlow, do you want to add something to it?

NOEL SWERDLOW: This is not a planetary observation. This is the 1515 observation of Spica. Copernicus says that he finds the longitude to be 17°14′ following from a declination of 8°36′ and a latitude of −2°. He uses Werner's value for the obliquity of the ecliptic of 23°28½′. If you compute the longitude with these parameters, the value will be 17°10′. You look at the manuscript and you discover that the original value written there is 17°10′. He then changed 17°10′ to 17°14′ to agree with the position that was computed from the precession tables after the theory was done.

JERZY DOBRZYCKI: But then he had to deal with the theoretical value of the obliquity.

NOEL SWERDLOW: He uses Werner's value for both observations of Spica (1515 and 1525) and his own value of 23°28⅖′ is merely computed from the tables. What I am saying is, that he goes from his theory back to the observations, and changes the observations to agree with the theory.

JERZY DOBRZYCKI: In some instances there is evidence in the manuscript that these corrections have been made during the writing of the chapter and appear to show that the corrections have been made on the basis of recomputation which was already done before the chapter was written. The parameters which follow have been written from the final values and these are found, with their corrections, on the previous page.

OWEN GINGERICH: I think this is also part of the whole spirit of handling observations before anything was known about error theory. I think one can site cases on this general sort in Ptolemy as well, where you have to work to get consistency from observations exceeding the number of parameters. Therefore, in the end some of the observations may get fudged.

EDWARD ROSEN: I shall be very brief in calling your attention to an inference that has been drawn from Copernicus' repeated attempts to arrive at a better approximate timing of an exact opposition. There is a German writer, who published a book entitled *Entstehung und Ausbreitung der Copernicanischen Lehre* in the middle of World War II, and a reprint of that volume is now on its way to reach us soon. In the course of this work Zinner accuses Copernicus of two things: falsifying data and being inaccurate in his computations.

This opinion, especially the deliberate falsification of data, has been widely reprinted from Zinner's adverse judgment. Now how does Zinner propose to cure this defect? He takes a particular instance and claims that the Moon could not have been used as an intermediary body for this particular observation since it was then only 25 hours old and therefore not visible. So he says that Copernicus evidently made an error of a whole month, and the date of the observation of October 10 should be made November 10, when the Moon was visible ($2\frac{1}{2}$ days old) and would have been available as an intermediary body. However, this great and learned scholar failed to notice that in the very next passage Copernicus computes the interval between two observations—year, month, day, hour, and fraction of hour—and if you recompute this interval you will find that Copernicus unquestionably meant October 10. He neither falsified his data nor did he make an error. What he was trying to do, and I think Professor Swerdlow is entirely right here, is to bring a crude observation, once made, into better conformity with the theory erected provisionally on the basis of that first approximation.

This is what we used to call rounding out when I was a student, smoothing the graph. When the dots did not all come on a smooth curve, we made a smooth curve that came fairly close to all the dots. And I believe that this is the correct understanding of what Copernicus did.

SESSION 2

The Humanistic Significance of Copernicus

Copernicus and the Quest for Certitude: "East" and "West"†

BENJAMIN NELSON

New School for Social Research, New York

I. E. Grant, P. Duhem, and Paolo Foscarini on Copernicus

No words that I can command can convey the story I have chosen to tell today more quickly and clearly than a confrontation of four passages from four authors, two writing in the seventeenth century and two speaking from our own time—two Western and two so-called Eastern. All four passages offer exceptional insight into the stakes at issue in the struggles over Copernicanism, anti-Copernicanism and the quest for certitude:

1. The first author is a young American scholar of our own day, Edward Grant of the University of Indiana, who is offering a summary of the meaning of Copernicus against the background of prior ways of discriminating astronomy and physics and establishing the criteria relevant for argument and proof in these spheres. Grant writes:

> Though Newton would start from the phenomena, his ultimate aim was to arrive at true laws which underlie the phenomena and by means of which "the things themselves are formed." Newton was searching for fundamental truths about the ultimate structure of matter from which its properties could be deduced. *This is in keeping with the great quest for reality initiated by Copernicus.*[1] [My italics.]

† For earlier discussions of the backgrounds of the issues to be dealt with in this paper, see Nelson (1965, 1967, 1973).

[1] Grant (1962), pp. 197–220. An essay of mine now in press will deal with predecessors of Copernicus in the "quest for reality" and "certitude"; see Nelson (1973–4).

> . . . modern science has shown a greater affinity with the XIVth century than with the century of Galileo and Newton. In the judgment of Pierre Duhem medieval scholastics had a truer conception of science than did most of the great scientists of the Scientific Revolution. Duhem even saw the Parisian nominalists as Christian positivists—forerunners of the positivist movement of his own day. He could not hide his scorn for the naïveté of some of the greatest figures of XVIIth-century science who confidently believed they could—and should—grasp and lay bare reality itself. Most of their basic errors, Duhem insisted, derived from their delusive search for reality which served only to corrupt the theoretical structure of science.
>
> Duhem is, in general, quite right. Scholastics were most sophisticated and mature in their understanding of the rôle which an hypothesis must play in the fabric of science. They were not, as we have seen, deluded into believing that they could acquire indubitable truths about physical reality. But it is an historical fact that the Scientific Revolution occurred in the XVIIth-century—not in the Middle Ages under nominalist auspices. Despite the significant achievements of medieval science—which Duhem himself did so much to reveal—it is doubtful that a scientific revolution could have occurred within a tradition which came to emphasize uncertainty, probability, and possibility, rather than certainty, exactness and faith that fundamental physical truth—which could not be otherwise—were attainable. *It was Copernicus who, by an illogical move, first mapped the new path and inspired the Scientific Revolution by bequeathing to it his own ardent desire for knowledge of physical realities.*[2] [My italics.]

2. The second writer is Paolo Foscarini, the Carmelite, whose Letter to the General of his Order was treated with greater harshness in 1616 than the work of Copernicus himself. The Decree of the Congregation of the Index forbade Foscarini's Letter absolutely. Foscarini writes as if animated by a desire to make clear the issue left in an ambiguous state by Grant.

Foscarini answers arguments of Catholics and Protestants alike by making the distinctions which proved of central importance in the ongoing disputes:

(a) "whilst otherwise the opinion (of Copernicus) is favored with *much probability* . . .".[3]

(b) *In reality* it is found "to be extremely repugnant (as at first sight it seems) not only to physical reasons and common principles received on all hands (which cannot do so much harm), but also (which would be of far worse consequence) to many authorities of Sacred Scripture. Upon which account many at first looking into it explode as the most fond paradox and monstrous caprice that ever was heard of."

(c) "Yet modern authors", Foscarini adds, "are induced to follow it, but with much hesitancy and fear, in regard that it seemeth in their opinion so to

[2] Grant (1962), pp. 219–20.

[3] The summary in the above paragraphs in the texts is drawn from Stimson (1917), pp. 71–72. On hearing my paper, Professor William Shea of the University of Ontario sensed that I would deeply appreciate having direct access to the sources on Foscarini he had gathered. I wish to acknowledge his great courtesy in sending me copies of the following critical materials: Foscarini's *Lettura*, the "*Judicium de epistola F. Pauli Foscarini de mobilitate Terrae*" and "*Defensio Epistolae F. Pauli Antonii Foscarini Veneti Ordinis Carmelitarum Teologi et Provincialis Provinciae Calabriae super mobilitate Terrae*". I now have a paper in progress which discusses these materials in detail.

contradict the Holy Scriptures that it cannot possibly be reconciled to them."

(d) Consequently—I follow Dorothy Stimson's words here—"Foscarini argued that the theory is either true or false; if false, it ought not to be divulged; if true, the authority of the Sacred Scriptures will not oppose it; neither does one truth contradict another. So he turned to the Bible."[4]

II. Matteo Ricci, Galileo, and Joseph Needham

The second batch of passages carries us to China. First, I focus on the most critical passages which Joseph Needham has written by way of setting forth his immense challenge to all interested in the comparative history, psychology and sociology of cultures, sciences, and civilizations—above all, to all interested in the so-called "grand titrations".

3. It will be recalled that in Needham's view China far outstripped the West in science and technology until the era between Leonardo and Galileo. It was, indeed, only with Galileo that a quantum jump occurred, the emergence of a distinctively new modern science, not simply a Western science, as he puts it, but a universal science. In this connection, Needham writes:

> It is vital today that the world should recognise that 17th century Europe did not give rise to essentially "European" or "Western" science, *but to universally valid world science, that is to say, "modern" science as opposed to the ancient and medieval sciences.* Now these last bore indelibly an ethnic image and superscription. Their theories, more or less primitive in type, were culture-rooted, and could find no common medium of expression. *But when once the basic technique of discovery had itself been discovered, once the full method of scientific investigation of Nature had been understood, the sciences assumed the absolute universality of mathematics, and in their modern form are at home under any meridian, the common light and inheritance of every race and people.* Of argument about elements and humours, Yin and Yang, or "philosophical sulphur," there could be no end, the disputants could reach no common ground. But the mathematisation of hypotheses led to a universal language, *an oecumenical medium of exchange*, a reincarnation of the merchants' single-value standard on a plane transcending merchandise. *And what this language communicates is a body of incontestable scientific truth acceptable to all men everywhere.* Without it plagues are not checked, and aircraft will not fly. The physically unified world of our own time has indeed been brought into being by something that happened historically in Europe, but no man can be restrained from following the path of Galileo and Vesalius, and the period of *political dominance* which modern technology granted to *Europeans is now demonstrably ending.*[5] [Italics mine.]

[4] Stimson (1917), pp. 71–72. Actually, Foscarini writes: " . . . O questa opinione de'Pittagorici è vera, ò nò; se non è vera non è degna, che si metta in campo: Se è vera, poco importa, che contradica à tutte i Filosofi, e gli Astrològi del mondo, e che per seguirla, e pratticarla s'habbia da fare una nuoua Filosofia, & Astrologia dependente da i nuoui principii, & hipothesi, che questa pone. Quello che appartiene alle scritture sacre, ne anco gli nuocerà, percioche una veritá non è contraria ali'altra; Se dunque è vera l'opinione Pittagorica, senza dubbio Iddio haurà talamente dettate le quell'opinione, e conciliamento con essa. . . ." (P. Foscarini, *Lettura* . . . (1615), pp. 12–13.)

[5] Needham (1954), esp. vol. III (1959), pp. 448–9.

4. The second passage relating to China is by an Italian Jesuit, Matteo Ricci, who is describing his reasons for undertaking a translation of Euclid into Chinese from the edition of his teacher, Christopher Clavius, S.J. It will be recalled that Ricci was the principal figure in one of the greatest Jesuit missions in the history of Christendom, that which brought a number of very distinguished Jesuit scientists, astronomers and engineers to China at an important turn in its history, the passage from the Ming to the Manchu, the time, also, of ardent efforts to reform the Chinese calendar. Ricci's journal is an exceptional source for certain aspects of this mission. Ricci writes:

> Shortly after they were settled down in the new residence, they went on an excursion through the villages of the Pekin Province, about three days' travel from the Capital. Their purpose in making this journey was to find out if the faith was spreading through the country, and they were not at all disappointed with what they discovered. Within the first years of their coming to Pekin, more than a hundred and fifty converts had come into the Church in this district and the numbers continued to increase with the years.
>
> It was during this time that the Fathers undertook a work which at first sight might not seem to be wholly in keeping with the purpose of their mission, but once put into practice proved to be quite beneficial. Doctor Ciu Paul had this one idea in mind: since volumes on faith and morals had already been printed, they should now print something on European sciences, as an introduction to further study, in which novelty should vie with proof. And so, this was done, but nothing pleased the Chinese as much as the volume on the Elements of Euclid. This perhaps was due to the fact that no people esteem mathematics as highly as the Chinese, despite their method of teaching, in which they propose all kinds of propositions but without demonstrations. The result of such a system is that anyone is free to exercise his wildest imagination relative to mathematics, without offering a definite proof of anything. In Euclid, on the contrary, they recognized something different, namely, propositions presented in order and so definitely proven that even the most obstinate could not deny them.[6]

III. Galileo, Copernicanism, and the Jesuits in China

In recent years a large discussion has developed as to the struggle over Copernicanism in the Far East. As many now know, it was the Jesuits, particularly Matteo Ricci, who was aware of the very great opportunities which existed for the Church to spread its mission among informed Chinese by the advancement of science in China. Pasquale D'Elia (S.J.),[7] Duyvendak,[8] and—above all—Joseph Needham[9] and most recently Nathan Sivin[10] have thrown immense light on the history of the Jesuit mission in China. The results seem to be as follows:

Thanks to Ricci some of the ablest Jesuits including associates of Galileo in the Academy of the Lynxes were sent to China to assist in many scientific activities, especially those associated with the improvement of the calendar and the fixing of

[6] Ricci, trans. L. J. Gallagher, S.J. (1953), pp. 476–7.
[7] D'Elia (1960).
[8] Duyvendak (1948), pp. 321–9.
[9] Needham (1954, 1970).
[10] Sivin (1973).

auspices, auguries and so on. The men who came included Adam Schall von Ball, Ferdinand Verbiest and others. Through the years the Jesuits maintained a continuing interest in the development of astronomy and physical theory in the West. Indeed, they sought to have Galileo's help and wrote letters asking about his specific findings on some astronomical issues. However, they were committed as an Order to serving the Pope and wholly refrained from adopting any position in favor of the Copernican view of Galileo.

The sentence against Galileo was honored without reservation, there were no explicit references to the Copernican hypothesis in works by Jesuit astronomers or by their Chinese associates for a long time to come. In China, as in Europe, the system which came to prevail was the Tychonic system which assumed the stability of the Earth and improved upon the Ptolemaic system for the rest of the Universe.[11]

Already several writers are insisting that Chinese science was retarded several centuries by the ambivalent reticences of the Jesuits in respect to the fortunes of Copernicus and Galileo. As everyone knows, the Jesuits were inhibited from teaching the Copernican hypothesis by the Decree of the Congregation of Index of 1616 and the injunction imposed on Galileo by Cardinal Bellarmine. Needham, in fact, insists that were it not for the Jesuits, the Chinese could have forged forward and very possibly achieved scientific advances equal to those of Galileo and Newton. In brief, Needham appears to say at times that it was the Jesuits who delayed the scientific revolution in China.

In view of Needham, the only interest the Jesuits originally had in seeking to relate the Chinese to Western science was to prove that Western religion with which Western science was associated was superior to the religion of China. Needham writes:

> The Jesuits' aim was naturally to support and commend the "Western" religion by the prestige of the science from the West which accompanied it. This new science might be true, but for the missionaries what mattered just as much was that it had originated in Christendom. The implicit logic was that only Christendom could have produced it. Every correct eclipse prediction was thus an indirect demonstration of the truth of Christian theology. The *non sequitur* was that a unique historical circumstance (the rise of modern science in a civilization with a particular religion) cannot prove a necessary concomitance. Religion was not the only feature in which Europe differed from Asia. But the Chinese were acute enough to see through all this from the very beginning. The Jesuits might insist that Renaissance natural science was primarily "Western", but the Chinese understood clearly that it was primarily "new".[12]

Duyvendak puts the matter with a somewhat different accent:

> The impossibility for the Jesuits, the mediators of Western science in China, to accept Galileo's heliocentric theory, is a matter of immense cultural significance. It meant that China, when it received Western science, received it in a form that, in principle, was already antiquated. It was cheated out of the most startling discovery that man had yet made, a

[11] *Ibid.*
[12] Needham (1954), esp. vol. III (1959), p. 449.

> discovery which, happily followed up as it was by men like Newton, may be regarded as the starting point of the phenomenal development in science and the revolution of man's thought that characterized the following centuries in Europe. "Modern" science in China remained essentially medieval; never were Chinese scholars faced with the appalling vision of the "T'ien-hsia" flying through space as a mere speck of dust.
>
> In Japan it was not otherwise. As late as 1795 the *Li-hsiang k'ao-ch'eng* . . . presumably in the revised edition of 1738, with additions by the missionaries, a book in which the sun is still supposed to revolve round the earth, was introduced in Japan as "modern" science. It is really with the condemnation of Galileo that the paths of East and West diverge, not to meet until the nineteenth century. . . .[13]

Hopefully, we will soon approach the day when we will be having full-scale documentary investigations of the intercivilizational outcomes of the decree against Copernicanism and the injunction and sentence against Galileo. Some of the newer work in these areas will be mentioned in our closing paragraphs below.

IV. Retrospect

It has been a major aim of this paper to show that from the start the struggle for and against Copernicanism in the "West" and "East" alike were powerfully caught up with battles over the entire structure of credibility and of the grounds of proof and evidence in all spheres of thought and, indeed, action. The battles which raged among those who accepted one or another variant or mix of fictionalism and probabilism and those who quested for objective certainty and subjective certitude were not confined to the spheres of astronomy and physics. They extended through every range of theory and practice, conduct and conscience.[14] The four passages cited in our first section from Edward Grant, Paolo Foscarini, Joseph Needham, and Matteo Ricci illustrate this clearly.

I have elsewhere contended that originally, at least in the twelfth and thirteenth centuries, probabilist notions quite evidently had the function of giving fresh options to those who did not wish to be under the dominion of proclaimed dogma in all spheres of faith and philosophy. As time passed, however, this very probabilist approach came to be the support and stay of fideism. Immense allowances existed for the proliferation of a great number of conjectures and opinions resting upon varying grades and shades of evidence, claiming varying measures of believability and credibility, enjoying so-called extrinsic as well as intrinsic probability.

It was not, as is so often supposed, the Church hierarchy which opposed probability. To the contrary the Church hierarchy was committed to computational fictionalism, the clash of probabilist conjectures and contrary-to-fact suppositions in natural philosophy; and to systems of equal and lesser probabilism in moral philosophy. One insufficiently stressed reason for the Church's favor of these

[13] Duyvendak (1948), p. 328.

[14] See again Nelson (1965, 1967).

perspectives was that these schemas did reserve to the Church a superordinate authority to define the truth. Only those who possessed the appropriate credentials were able to claim the right to interpret the Book of Revelation and, therefore, the Book of Nature.[15]

As Nathan Sivin illustrates in a recent study, the battles over probability and credibility persisted in China into the nineteenth century.

There are yet other reasons for the long delay of the adoption of the new system of the world in the Far East. Editorial limitations do not follow us here to enter into these issues in the detail they deserve, but the report by Matteo Ricci cited above as well as other evidences which are in clear focus in Chinese and Japanese sources make it clear that great breakthroughs needed to occur at every level of differentiation in the social and cultural spheres before new openings could be made to scientific universalism and universality.[16] If Ricci stresses the lack of clear, unambiguous notions of proof, Nakayama tells us a great deal about the powerfully institutionalized secrecies in the activities of Japanese scientists attached to Imperial departments and working in the Imperial service.[17]

It is hardly possible to exaggerate the importance of these wider social and cultural factors in the telling of the story of the spread of modern science across the world.

References

D'ELIA, P., S.J. (1960) *Galileo in China: Relations through the Roman College between Galileo and the Jesuit Scientist-Missionaries (1610–1640)*. Cambridge, Mass.: Harvard University Press.

DUYVENDAK, J. J. L. (1948) "P. D'Elia's *Galileo in China*." In *T'oung Pao*, 38, 321–9, esp. p. 328.

FOSCARINI, P. (1615) *Lettera sopra l'opinone de' Pittagorica e del Copernico della mobilita della terra*. Naples: per Lazaro Scoriggio.

GRANT, EDWARD C. (1962) "Late medieval thought, Copernicus and the scientific revolution." *Journal of the History of Ideas*, 23, 197–220.

JONES, CHRISTINE (1964) The Geoheliocentric Planetary System: Its Development and Influence in the Late Sixteenth and Seventeenth Centuries. Unpublished Ph.D. Dissertation, Cambridge University.

NAKAYAMA, SHIGERU (1969) *A History of Japanese Astronomy: Chinese Background and Western Impact* (Harvard-Yenching Institute Monograph Series, 18). Cambridge, Mass.: Harvard University Press.

NEEDHAM, JOSEPH (1954) *Science and Civilization in China* (4 vols. in 7 parts). New York: Cambridge University Press.

NEEDHAM, JOSEPH (1970) *The Grand Titration*. Toronto: University of Toronto Press.

NELSON, BENJAMIN (1965) " 'Probabilists,' 'Anti-Probabilists,' and the quest for certitude in the 16th and 17th centuries." *Actes du Xme Congrès internationale d'histoire des sciences*. Paris: Hermann, vol. 1, pp. 267–73. (Original draft, 1962.)

---

[15] Nelson (1973–4).

[16] Nelson (1973, 1973a).

[17] Nakayama (1969).

NELSON, BENJAMIN (1967) "The early modern revolution in science and philosophy: fictionalism, probabilism, fideism, and Catholic 'Prophetism'." *Boston Studies in the Philosophy of Science*, R. S. COHEN and M. WARTOFSKY (eds.). Dordrecht, Holland: D. Reidel, Vol. 3, pp. 1–40.

NELSON, BENJAMIN (1973) "On the shoulders of the giants of the *comparative* historical sociology of 'science'—in civilizational perspective." *Social Processes of Scientific Development*, R. WHITLEY (ed.). London: Routledge & Kegan Paul, pp. 13–20.

NELSON, BENJAMIN (1973a) "Civilizational complexes and intercivilizational encounters." *Sociological Analysis*, Vol. 34, pp. 79–105.

NELSON, BENJAMIN (1973–4) "The work of God's hand, the books of nature and revelation, and the machine of the world: 11th–17th centuries." Presented to a "Round Table" of the *Boston Colloquium for the Philosophy of Science*, April 3, 1973, Boston: Boston University.

RICCI, MATTEO (1953) *China in the Sixteenth Century: The Journals of Matthew Ricci: 1583–1610.* Tr. from Latin by L. J. GALLAGHER, S.J. New York: Random House.

SIVIN, NATHAN (1973) "Copernicus in China." *Colloquia Copernicana, II* (Conférences du Symposium de l'UIHPS. Toruń, 1973). Wrocław-Warszawa . . .: Polskiej Akademii Nauk, 1973, pp. 63–114.

STIMSON, DOROTHY (1917) *The Gradual Acceptance of the Copernican Hypothesis.* Ph.D. Thesis. Columbia University. Pub. by the author, Hanover, N.H.

Some Episodes in the History of Copernicanism

Robert Palter

University of Texas at Austin, Austin, Texas

Like many celebrated thinkers—and perhaps more than most—Copernicus has been the victim of persistent misunderstanding by his successors. Already with the publication of his *magnum opus* Copernicus suffered the indignity of being introduced to his readers by an anonymous preface which he not only never saw but which deliberately misrepresented his views on the nature of astronomical hypotheses. And if Copernicus' great advocate, Kepler, was later to reveal publicly for the first time the identity of the author of this preface, Copernicus' other great advocate, Galileo, was responsible for popularizing several glaring errors concerning Copernicus' career (though, as Professor Rosen assures us, these "should be considered honest mistakes rather than conscious falsehoods").[1] Perhaps in the heat of the battle over the competing world-systems during the seventeenth century, it would have been unrealistic to expect too much in the way of accuracy, objectivity, and fairmindedness. Unfortunately, later commentators, with far less justification, have also rarely succeeded in evaluating Copernicus without slipping into hagiolatry or some other form of special pleading. Not to conceal my own views, let me just say that for me Copernicus is no sacrosanct saint of science (any more than he was a priest of the Catholic Church, as Galileo erroneously claimed). But neither was he simply a dull, uninspired traditionalist, content to perpetuate certain already outmoded astronomical methods. If it be true that without Kepler nothing worth calling an

[1] E. Rosen, "Galileo's misstatements about Copernicus", *Isis*, 49, 3 (Sept. 1958), 330.

astronomical revolution would have occurred at all—and this is today perhaps the most common view among historians of astronomy—it may also be true that without Copernicus there would have been no Kepler. This line of thought I shall pursue no further here, however, because my immediate concern is with methodological clarity rather than with historical details.

The question I wish to raise is in a way pre-methodological: not so much "What are the appropriate methods for studying Copernicanism?" as "What are the main general types of questions that might appropriately be raised in the course of such Copernican studies?" Among the questions of this latter kind, I consider the following three to be exemplary: (1) What does Copernicus *actually say and mean* in the *Revolutions* (or in any one of his other astronomical writings)? (2) What are the *fundamental conceptions* of Copernicus' astronomical system (or systems)? (3) How are Copernicus' astronomical writings *related* to earlier, later and contemporary writings on astronomy and on other subjects (such as physics, astrology and philosophy)? These three questions suggest three main approaches to the study of Copernicanism, which I shall characterize as *episodic*, *conceptual* and *relational.* Though I believe this three-fold distinction applies to historical studies generally, in view of the subject at hand I shall restrict my explication of the distinction to the case of Copernican studies.

To study the text of Copernicus' *Revolutions* with a view to understanding, in its author's own terms, what is said and why it is said, the mixture of truth and error, insight and oversight—all this is to be engaged in *episodic* analysis of one concrete and fully determinate version of Copernicus' astronomical system. (Other versions might conceivably be found in other writings by Copernicus.) To formulate some more or less reconstructed model of the astronomical system of Copernicus' *Revolutions* with a view to understanding, in abstract general terms, certain aspects of that system's fundamental structure, some of its potentialities and limitations as an astronomical theory, some of its logical interrelations with other scientific theories—all this is to be engaged in *conceptual* analysis of Copernican astronomy (not to be confused with analysis of some actual version of Copernicus' astronomical system). Finally, to study the antecedents, consequents, or contemporaries of Copernicus' *Revolutions* with a view to understanding, in cultural or sociological terms, its genesis, its subsequent interpretation, distortion and transformation in a variety of scientific, pedagogical and philosophical contexts, its place in the total historical context of its time—all this is to be engaged in *relational* analysis of Copernicus' *Revolutions* (in which enterprise Copernican models will normally be indispensable analytical devices).

It should be noted that episodic analysis deals in the first instance with *historical particulars* (e.g. the publications of specific written texts); conceptual analysis deals in the first instance with *trans-historical abstractions* (e.g. Copernicus' astronomical system shorn of certain "accidental" features, such as the fictitious variations in the rate of precession and in the obliquity of the ecliptic); relational analysis deals in the first instance with *configurations of historical particulars*, whether successive (e.g. Ptolemy's *Almagest*, Copernicus' *Revolutions*, and Galileo's *Dialogue*) or synchronic (e.g. the publication of Kepler's *New Astronomy*, and Galileo's first telescopic discoveries, each of which occurred in the year 1609). Ontologically, these particulars, abstractions and configurations are mutually dependent and interacting; likewise, the three corresponding types of historical analysis are closely interrelated. Nevertheless, there is a certain ontological priority associated with the particulars, and this is reflected in what I take to be the primacy of the episodic approach in studying history. (It should be clear, incidentally, that every instance of historical analysis—whether episodic, conceptual, or relational—itself constitutes an historical episode and as such may be subjected to historical analysis of any one of the three types.) Many historical studies adopt all three approaches simultaneously. But while this may make for lively and provocative historical writing, it also runs the risk of dangerous ambiguity and confusion. To illustrate from Copernican studies: the question of what exactly Copernicus wrote and believed is very different from the question of what is taken for Copernicanism by later astronomers, theologians, historians or poets; and both of these questions obviously differ from the question of whether in some fundamental mathematical sense geokinetic astronomy is "simpler" than heliokinetic astronomy. The scholarly world could be spared much pointless and distracting controversy if the varying status of questions of this kind were always kept clearly in mind.

I turn now to some episodes in the history of Copernicanism, by which I mean certain discrete and well-defined historical events involving one or another geokinetic model. In most cases, the episode's dominant model will consist simply in the conception of a rotating Earth revolving about the Sun. It may be worth noting, however, that the Sun's motion differs from one Copernican model to another: thus, for Copernicus himself the Sun is at rest; for Kepler and Galileo the Sun rotates but remains in a fixed position; for Newton the Sun not only rotates but also moves in a small orbit about the center of gravity of the solar system. Whether one chooses to characterize these as different models or merely as variants of one basic model can only be decided in terms of the direction and purpose of the particular historical analysis in hand.

My first episode is the engraving of a map of the world, possibly by Hans Holbein the Younger (Fig. 44, p. 60). At each of the Earth's poles the artist has depicted a winged figure turning a crank to suggest the diurnal rotation. (Analogous depictions of angels rotating the celestial sphere occur in both medieval and Renaissance representations of the Ptolemaic universe; e.g. Fig. 45, p. 61.) Since the map was included in some copies of Simon Grynaeus (or Johann Huttich), *Novus Orbis*, published by Hervagius in Basle in 1532, this episode may well constitute the earliest printed allusion to the Earth's rotation. Where the artist obtained the idea of a rotating Earth remains an open question. One clue may lie in the fact that a tract by Calcagnini on the rotation of the Earth was first published, in 1544, by Froben, the publisher with whom Holbein was closely associated. (Calcagnini died in 1541 but had completed the tract at least 20 years before its publication.) Relational analysis of this Copernican episode to detect possible influences from Copernicus himself would be of great interest.[2]

My second episode consists of a single remark by Pascal. Among his *pensées* is one which includes this passage: "I approve of not examining the opinion of Copernicus." This apparent refusal by one of the greatest scientists of the seventeenth century to take a stand on the crucial cosmological debate of his time has shocked many of his modern commentators. My own discussion of the passage will follow that of Allix in his excellent paper of 1904.[3]

To understand Pascal's attitude toward Copernicanism, it is necessary to examine first the entire *pensées* from which the offending passage was excerpted. I quote the whole of *pensée* 218: "*Dungeon* [cachot] . . . I approve of not examining the opinion of Copernicus; but this . . .! It concerns all our life to know whether the soul be mortal or immortal."[4] It is immediately evident that the context is religious, and this suggests that we shall eventually have to shift from episodic to relational analysis in order to appreciate fully what Pascal is saying. But before doing this, let us take note of another Pascalian text, written probably at about the same time as the previous one. In the last of his *Provincial Letters*, written in March 1657, Pascal refers to Galileo's condemnation by the Inquisition some 24 years earlier in the following terms:

> It was to equally little purpose that you obtained against Galileo a decree from Rome, condemning his opinion respecting the motion of the earth. It will never be proved by such an argument as this that the earth remains stationary; and if it can be demonstrated by sure observation that it is the earth and not the sun that revolves, the efforts and arguments of all

[2] I owe my acquaintance with this map to W. Oakeshott, who ascribes it, with very little supporting evidence, to Holbein ("Some classical and medieval ideas in Renaissance cosmography", *Fritz Saxl, 1890–1948*, ed. D. J. Gordon, pp. 253–4, 258–60 [London, 1957]). Oakeshott thinks that Holbein probably executed the engraving prior to 1530 (which would have been sometime before his departure from Basle in 1532 for his final sojourn in England). Oakeshott reproduces (in his plate 21) an illustration of angels turning the celestial sphere, though without cranks; a fourteenth-century illustration in which the angels use cranks may be found reproduced in Alan G. R. Smith, *Science and Society in the Sixteenth and Seventeenth Centuries*, p. 10 (London, 1972). The map is also reproduced in L. Bagrow, *History of Cartography* (ed. R. A. Skelton, p. 85, fig. 19 [Cambridge, Mass., 1964]), where it is attributed to Sebastian Münster. In view of Münster's Ptolemaic astronomical views—see Fig. 45—this attribution does not seem too likely. It should be noted, however, that the *Novus Orbis* does contain a 12-page disquisition by Münster on a world map. The complicated bibliographical details concerning the authorship and publication of both the map and the book in which it was included may be studied in two works by Henry Harrisse: *Bibliotheca Americana Vetustissima*, pp. 291–6 (New York, 1866), and *The Discovery of North America*, pp. 586–8 (London–Paris, 1892; reprinted, Amsterdam, 1961).

[3] M. G. Allix, "Pascal et le Système de Copernic", *Bulletin de l'Académie Delphinale*, xviii, 267–95 (1904).

[4] Pascal, *Pensées*, 218, trans. W. Trotter, p. 76 (New York, 1941).

> mankind put together will not hinder our planet from revolving, nor hinder themselves from revolving along with her.[5]

The motion of the Earth is introduced as an example of a matter of fact which can only be known by observation rather than by either faith or reason. It is thus clear that no fear of ecclesiastical disapproval entered into Pascal's attitude toward Copernicanism. Why, then, did he recommend a non-committal stance? Apparently—and here we move to the larger cultural context of Pascal's remark—in order to differentiate *scientific* scepticism from the more thoroughgoing scepticism extending to *all* knowledge-claims of his great predecessor and fellow-countryman, Montaigne.

To document this interpretation Allix cites numerous passages from Pascal's writings, especially the *pensées*, in which Montaigne is singled out for criticism on account of the "absolutely pernicious" effect of his scepticism on those with any prior tendency toward impiety. Thus, in one place, Pascal writes of Montaigne that "one cannot excuse his thoroughly pagan views on death, for a man must renounce piety altogether, if he does not at least wish to die like a Christian".[6] Furthermore, the very allusion to a dungeon in *pensée* 218 may be traced back to Montaigne, as may be seen by comparing a passage in the great *pensée* 72 on "Man's disproportion" with a passage in the "Apology for Raimonde Sebonde".

> . . . let [man] regard himself as lost in this remote corner of nature; and from the little cell [cachot] in which he finds himself lodged, I mean the universe, let him estimate at their true value the earth, kingdoms, cities, and himself.[7]
>
> You see only the order and government of this little cave [caveau] you dwell in, at least if you do see it. His divinity has infinite jurisdiction beyond; this part is nothing in comparison with the whole.[8]

Finally, among the immense number of items which Montaigne catalogues in support of scepticism and fideism in the "Apology for Raimonde Sebonde" is the topic of Copernicanism:

> The sky and the stars have been moving for three thousand years; everybody had so believed, until it occurred to Cleanthes of Samos, or (according to Theophrastus) to Nicetas of Syracuse, to maintain that it was the earth that moved, through the oblique circle of the Zodiac, turning about its axis; and in our day Copernicus has grounded this doctrine so well that he uses it very systematically for all astronomical deductions. What are we to get out of that, unless that we should not bother which of the two is so? And who knows whether a third opinion, a thousand years from now, will not overthrow the preceding two?[9]

---

[5] Pascal, *The Provincial Letters*, Letter XVIII, trans. T. M'Crie, p. 615 (N.Y., 1941).

[6] Pascal, *Pensées*, 63, p. 20.

[7] *Ibid.* 72, p. 22.

[8] *The Complete Essays of Montaigne*, trans. D. Frame, p. 389 (Stanford, Calif., 1958).

[9] *Ibid.*, p. 429. "Cleanthes of Samos" should, of course, be "Aristarchus of Samos". In Pierre Michel's annotated edition of Montaigne's *Essais* (III, p. 341, n. 523 [Paris, 1957]) the editor claims that Montaigne derived this error from Plutarch's essay "Concerning the Face Which Appears in the Orb of the Moon". But in fact there is no such confusion in Plutarch, who wrote: "Cleanthes thought that the Greeks ought to lay an action for impiety against Aristarchus the Samian" (*Plutarch's Moralia*, XII, 923A, trans. H. Cherniss and W. Helmbold, p. 55 [Cambridge, Mass., 1957]).

Accepting, indeed, paraphrasing Montaigne's view of the cosmological dispute, Pascal also uses the triviality of the entire question, as he sees it, as a foil for the truly important question of the immortality of the soul. As a scientific thinker, Pascal no doubt appreciated the arguments in favor of Copernicanism; as a religious man, he had to face the utter unimportance of all such scientific issues in the light of uncertainty about the burning issue of eternal life.

In keeping with the chastened hopes and diminished prospects of twentieth-century man, Albert Camus expresses precisely the same attitude toward Copernicanism as Montaigne and Pascal, but the truly important question now becomes not one of eternal life but one of any life at all, in short, the question of suicide. Thus, Camus opens his essay "An Absurd Reasoning" with the proposition that "There is but one truly serious philosophical problem, and that is suicide",[10] and then proceeds to argue that no purely theoretical problem can have the urgency of the problem of "Judging whether life is or is not worth living".[11] As instances of such theoretical problems Camus mentions the ontological argument for the existence of God and the motion of the Earth, remarking about the latter that "Whether the earth or the sun revolves about the other is a matter of profound indifference. To tell the truth, it is a futile question."[12] We see that the language of scepticism has hardly changed since the sixteenth century. (Even some of the names of the sceptics remain the same: Montaigne had a disciple named Jean-Pierre Camus (1584–1654), whose *Essay Sceptique* uses the debate over Copernicanism in the by then conventional way as one more argument against the possibility of any genuine knowledge.)[13]

I turn now to some historical episodes in which Copernican models are evaluated, positively or negatively, rather than simply entertained as abstract possibilities or deployed as arguments for scepticism. It will clarify later discussion to begin with a few remarks about Copernicus' *Revolutions*. This is a long and complicated and, in some respects, a baffling book. Many questions about the book (e.g. its possible dependence on Arabic sources for certain technical geometrical innovations) remain unresolved by present-day scholars. There is, however, one non-controversial aspect of the book which cannot, I think, be overemphasized in assessing its historical fortunes. I refer to the striking contrast between Book I and the remaining five books of the work. This contrast becomes immediately apparent simply by turning the pages of a copy of the *Revolutions*: the tables of astronomical data and the geometrical diagrams which are characteristic of Books II–VI are entirely absent from Book I. Another way of putting this is to say that Book I is primarily concerned with physical astronomy (or cosmology)—in Copernicus' words "the general system of the

[10] A. Camus, *The Myth of Sisyphus*, trans. J. O'Brien, p. 3 (N.Y., 1955).

[11] *Ibid.*

[12] *Ibid.*

[13] On Jean-Pierre Camus, see R. Popkin, *The History of Scepticism*, rev. ed., pp. 63–66 (N.Y., 1968).

Universe"—and the remaining books primarily with mathematical astronomy.[14] In terms of their contents, Book I is designed to demonstrate in a general way the so-called three-fold motion of the Earth (and the corresponding central and static position of the Sun), while Books II–VI are designed to explain in full mathematical detail the observed motions of the planets (including, of course, the Earth and its moon). Granted, then, that the *Revolutions* is composed of these two distinct—though not unrelated—parts, one may ask two questions: first, how did Copernicus himself conceive of the relation between these two parts? and second, how did Copernicus' immediate successors react to and interpret these two parts? The latter question is relatively easy to answer; the former, not easy at all.

With the exception of Nikolaus Müller's carefully corrected and annotated 1617 edition—the third—of Copernicus' *Revolutions*, sixteenth- and seventeenth-century publications dealing with Copernican models tended to concentrate on one or the other of the two parts of his book. Thus, Erasmus Reinhold based his *Prutenic Tables* (1551) on the astronomical tables in the *Revolutions* but rejected the motion of the Earth;[15] John Feild recalculated the *Prutenic Tables* for the meridian of London, but his *Ephemeris*, published in 1556, took no stand on the motion of the Earth, not even in the preface, by Feild's mentor, John Dee;[16] whereas the numerous poets, theologians and cosmological writers who debated the relative merits of Ptolemy, Copernicus and Tycho Brahe usually ignored completely the technical details of mathematical astronomy. Even Galileo, in arguing for Copernicus and against Ptolemy, avoids introducing such details. Kepler was exceptional in his deep concern for both parts—the physical and the mathematical—of Copernicus' astronomical work. After Kepler, of course, Copernicus' mathematical astronomy became outmoded and could be safely ignored even by astronomers, though Copernican views on the motion of the Earth continued to be debated right up to the time of Newton.

But why was not Copernicus himself more sensitive to the interaction—and indeed to the tensions—between the two parts of his *magnum opus*? Why was he boldly innovative in Book I and timidly conservative or worse in the other books? Why did he fail so abjectly to understand the true advantages of the system he was proposing? Despite the popularity of such questions even among otherwise knowledgeable and sophisticated scholars, the tendentious rhetoric and misleadingly anachronous

[14] The phrase quoted may be found in Copernicus, *On the Revolutions of the Heavenly Spheres*, Preface, trans. J. Dobson and S. Brodetsky, *Theories of the Universe*, ed. M. Munitz, p. 151 (Glencoe, Ill., 1957). The distinction between physical and mathematical astronomy is discussed at length in my "An Approach to the History of Early Astronomy", *Studies in the History and Philosophy of Science*, I, 2, 93–133 (1970).

[15] See E. Rosen, *Three Copernican Treatises*, 2nd ed., p. 205 (N.Y. 1959).

[16] Dee's work in astronomy is discussed at length in Francis Johnson, *Astronomical Thought in Renaissance England* (Baltimore, 1937); Feild's *Ephemeris* is briefly noticed by Johnson on pp. 134–5. The current *Dictionary of Scientific Biography* (ed. C. Gillispie, vol. 4, 1972) includes no entry on Feild; the *DNB* has a brief account of his life and writings, where he is referred to as a "proto-Copernican".

character of such questions ought to be recognized for what they are. Brevity forbids any discussion of these matters beyond the following few sketchy remarks. Copernicus was mainly concerned to demonstrate the feasibility of a moving earth as a sound physical basis for astronomy and to this end he subordinated most of the geometrical details of his mathematical astronomy. For example, the physical principle that uniform circular motions are "natural" to spherical bodies—thereby justifying the circular motions of the Earth—leads to Copernicus' insistence on the primacy of uniform circular motion in his geometrical models for the motions of all heavenly bodies (the notorious rejection of Ptolemy's equant). On the other hand, to show that the traditional technical devices of heliokinetic astronomy (including deferents, epicycles and eccentrics) still worked even when the physical basis of astronomy was radically altered might well have seemed a particularly persuasive way of demonstrating the superiority of the new physical basis. This superiority Copernicus explicates as follows:

> . . . if the motions of the rest of the planets be brought into relation with the circulation of the Earth and be reckoned in proportion to the orbit of each planet, not only do their phenomena presently ensue, but the orders and magnitudes of all stars and spheres, nay the heavens themselves, become so bound together that nothing in any part thereof could be moved from its place without producing confusion of all the other parts and of the Universe as a whole.[17]

But why, it may be asked, did Copernicus have to undertake the immense labor of, in effect, reworking traditional mathematical astronomy when it should have been perfectly clear to him (as presumably it is to us) that all that was involved was a fairly trivial shift in point of reference?[18] Here I would merely suggest that in the absence of a general theory of such mathematical transformations in Copernicus' time, his results could not—and did not—appear trivial; and this in spite of Copernicus' clear formulation of a principle of relative motion.[19] What Copernicus did was to demonstrate once and for all that mathematical astronomy was fully compatible with a point of reference for celestial motions other than the Earth. Copernicus' failure to take full advantage of his shift in reference point—so as, for example, to improve the traditional theory of planetary latitudes—may not, after all, be so important when compared with his actual achievement.

One way of expressing the variety of response to Copernicus' astronomical system is to say that different thinkers have extracted different conceptual models, or

[17] Copernicus, *op. cit.*, p. 151.

[18] As O. Neugebauer puts it: "The choice of the reference system has no effect whatever on the structure of the model [of the astronomical system]" (*The Exact Sciences in Antiquity*, 2nd ed., p. 204 (Providence, 1957). Cf. O. Neugebauer, "On the planetary theory of Copernicus", *Vistas in Astronomy*, ed. A. Beer, **10**, p. 103 (Oxford, 1968): ". . . the Copernican theory is only a formal transformation of the Ptolemaic theory."

[19] Copernicus, *op. cit.* I, 5, p. 157: "A seeming change of place may come of movement either of object or of observer, or again of unequal movements of the two. . . ."

variants of a single model, from their study of the *Revolutions*. It is this process of conceptual reconstruction that I should now like to illustrate.

In his great book on relativity physics, *Space–Time–Matter*, Hermann Weyl begins his discussion of the relativity of space and time (Chapter III) by pointing to certain well-known facts which make plausible the relativity of rest and motion; namely, rest and motion in everyday life are always defined with respect to some set of reasonably permanent and stable bodies, such as buildings, ships and, ultimately, the crust of the Earth. Generalizing, we may say that "Only the motions of bodies (point-masses) relative to one another have an objective meaning".[20] To translate this principle into exact mathematical terms, we (following Weyl) first imagine each point-mass to be characterized by three (rectangular Cartesian) space coordinates x, y, z, and one time coordinate, t. Now, to assert the relativity of motion of point-masses to one another is the same as asserting the admissibility of any coordinate system moving arbitrarily with respect to the x, y, z, t coordinate system. At any given moment the relations between two rectangular Cartesian coordinate systems sharing a single (absolute) time coordinate may be expressed as follows:

$$\begin{aligned} x &= a_{11}x' + a_{12}y' + a_{13}z' + a_1 \\ y &= a_{21}x' + a_{22}y' + a_{23}z' + a_2 \\ z &= a_{31}x' + a_{32}y' + a_{33}z' + a_3 \\ t &= t' + a, \end{aligned}$$

where a, a_1, a_2, a_3, are arbitrary constants and the a_{ij} are constants satisfying the conditions for a so-called orthogonal transformation (conditions which follow directly from the rectangular Cartesian character of the two coordinate systems).[21] Since the relations between the two coordinate systems will in general change with time, a_1, a_2, a_3, and the a_{ij} must in fact be understood as continuous functions of the time t', such that for any fixed value of t' the a_{ij} are coefficients of an orthogonal transformation. (a represents simply the difference in time-origin in t and t', and cannot of course itself vary with time.)

As an example of such a coordinate transformation consider two coordinate systems, S: x, y, t, and E: x', y', t'. (One space coordinate is suppressed in the interest of simplicity.) Suppose that x' and y' are parallel respectively to x and y and that the origin of E rotates with uniform velocity ω_E about the origin of S in a circle of radius

[20] H. Weyl, *Space–Time–Matter*, 4th ed., trans. H. Brose, p. 152 (N.Y., 1922).

[21] The conditions may be expressed as follows:

$$a_{ij}a_{ik} = \delta_{jk} \qquad (j, k = 1, 2, 3),$$

where $\delta_{jk} = 0$ for $j \neq k$ and 1 for $j = k$. Precisely analogous conditions hold for the coefficients a'_{jk} of the inverse transformation from x, y, z, t to x', y', z', t'. Weyl goes on to show how these coordinate transformations may be restricted so as to express the so-called principle of Galilean relativity (*ibid*., p. 153).

r_E. We then have:

$$x = x' + r_E \cos \omega_E t',$$
$$y = y' + r_E \sin \omega_E t',$$
$$t = t' + a.$$

Let us now imagine that S and E represent coordinate systems attached to the Sun and the Earth respectively, and suppose we wish to describe the motion of an outer planet P, which rotates with uniform velocity ω_P in a circle of radius r_P about the origin of S. The equations of motion of P in the coordinate system S would then be:

$$x = r_P \cos \omega_P t,$$
$$y = r_P \sin \omega_P t,$$

from which one immediately derives:

$$x^2 + y^2 = r_P{}^2,$$

confirming the circular shape of the orbit. Transforming to the E-coordinates, one finds:

$$x' = r_P \cos \omega_P t' - r_E \cos \omega_E t',$$
$$y' = r_P \sin \omega_P t' - r_E \sin \omega_E t',$$

which show that the position of P is now a rather complicated function of the time and hence that the orbit of P is some rather complicated plane curve such that the time parameter cannot be readily eliminated from its two defining equations.

Let us return now to Weyl. It is, I believe, in view of considerations like those just presented, that he offers the following evaluation of Copernicus' achievement:

> This is the substance of the discovery of Copernicus that was afterwards elaborated to such an extraordinary degree by Kepler, namely, that there is in fact a co-ordinate system for which the laws of planetary motion assume a much simpler and more expressive form than if they are referred to a motionless earth.[22]

But while this judgment of Weyl's is true of the simple concentric circular motion model I have constructed, it is scarcely relevant to the actual historical systems of Copernicus and his rivals. Whatever illumination we may derive from such reconstructed Copernican models will do little to help us see sixteenth- and seventeenth-century astronomy in its true light. No astronomer in that period, after all, thought that any planet revolved uniformly in a circle about the Sun (or about the Earth, in the case of anti-Copernicans). Kepler's elliptical orbits are another matter, and Weyl is surely correct if he means to be pointing primarily to the enormous gain in conceptual simplicity which came with Kepler's replacement of combinations of circles by straightforward ellipses (though any comparable direct gain in practical computational simplicity or in physical plausibility is much more problematic).

[22] H. Weyl, *Space–Time–Matter*, 4th ed., trans. H. Brose, p. 152 (N.Y., 1922).

Obviously, Weyl's analysis can stand for an entire large class of similar analyses in which stress is laid on the alleged role of a criterion of simplicity in deciding between heliokinetic and geokinetic astronomy. Most such analyses known to me suffer fatally from a failure to distinguish sharply between episodic analysis of sixteenth- and seventeenth-century astronomy and conceptual analysis of the mathematical significance of coordinate transformations in describing the motions of heavenly bodies.

I turn next to certain Copernican episodes in which the central concern is with the possible role of sense-experience in confirming the motion of the Earth. The great seventeenth-century source for this topic is, of course, Galileo's *Dialogue on the Two Great World-Systems.* The Copernican model used throughout this work is a minor variant of that found in Book I of Copernicus' *Revolutions*: all three of Copernicus' terrestrial motions are introduced for discussion but the third—the conical motion of the Earth's axis—is dismissed as redundant in view of the assumption that the natural (we would say "inertial") motion of the Earth is one in which any given terrestrial diameter retains a constant orientation in space. The arguments advanced in favor of the rotational and orbital motions of the Earth are of two kinds: negative arguments which explain why these motions are not ordinarily observed, and positive arguments which purport to explain certain observed phenomena which result from these motions. The positive arguments are just two in number, namely, the existence and properties of the tides (to which the entire last part of the *Dialogue* is devoted) and the characteristic motions of sunspots (discussed in the third part of the *Dialogue*). The negative arguments are of two types: first, arguments based on a principle of relativity and leading to the conclusion that certain observed phenomena (such as the motion of projectiles) are consistent with the motion of the Earth; second, arguments leading to the conclusion that certain expected consequences of the motion of the Earth (such as stellar parallax and the existence of centrifugal force) are just too small to be observed.

Galileo emphasizes the inconclusive results of what I have called his negative arguments: not only is there no direct sensory confirmation of the motion of the Earth, but sensory experience strongly suggests the *absence* of any such motion. The Copernicans have, in Galileo's famous words, "through sheer force of intellect done such violence to their own senses as to prefer what reason told them over that which sensible experience plainly showed them to the contrary".[23] Lest it be thought that Galileo means here to be rejecting, or even slighting, the probative force of sense-experience in the quest for knowledge of nature, I hasten to add that the passage quoted occurs in the Third Day of the *Dialogue*, prior to consideration of both sunspots and the tides, the two phenomena which are introduced precisely to provide definite empirical evidence for the motion of the Earth. Galileo's considered position

[23] Galileo, *Dialogue Concerning the Two Chief World Systems*, trans. S. Drake, p. 328 (Berkeley, 1962).

is, I would judge, that mere unaided observation by means of the senses is incapable of confirming the moving-earth hypothesis—and may indeed suggest the very opposite hypothesis—but that appropriate theoretical, or rational, analysis enables one to derive observationally confirmable consequences from the moving-earth hypothesis.

Theories of knowledge different from Galileo's have, predictably, led to other interpretations of the moving-earth hypothesis. Some recent episodes in the history of Copernicanism have challenged Galileo's conviction that the adherents of the fixed-earth hypothesis could be persuaded of the errors of their ways by empirical evidence (such as that provided by the phenomenon of the tides). Thus, the late Norwood Hanson argued that Tycho Brahe and Kepler did not see the same thing in the sky at dawn and dusk, Tycho observing the Sun in motion on each occasion and Kepler observing the horizon first falling away from and then approaching the Sun.[24] Concerned to illustrate his thesis about the relativity of observational data to theories or conceptual frameworks, Hanson attempts to trade on a very dubious analogy drawn from experimental psychology: observing the "Sun rise" or the "horizon fall" is compared with the Gestalt shift which occurs as one observes certain specially drawn figures (e.g. Necker perspective cubes). But the Gestalt shifts are involuntary, whereas geokinetic or heliokinetic frameworks are freely adopted by astronomers after due deliberation. And while it is not uncommon to use the verb "see" in Hanson's sense of "see that" (or "know that" or "believe that"), it is by no means clear that even this fact will save Hanson's argument.[25]

To accept the Copernican model is, implicitly at least, to accept the condemnation of what one sees in the sky at sunrise and sunset as mere appearances, but this in no way alters what one does in fact see. Indeed, I would suggest that under normal conditions one cannot see the horizon moving at all. This is because under normal conditions it is precisely the horizon which provides the fixed framework relative to which all moving objects are observed. In the words of James J. Gibson (an outstanding authority on perception of motion): "The phenomenal motion of objects is not ordinarily perceived as relative to an arbitrary frame of reference. The stable visual world provides an absolute zero and hence there is an absolute sense of motion or rest with reference to the ground—both for oneself and objects."[26] Of course I do not wish to deny that under abnormal conditions (as when the observer himself is

[24] N. Hanson, *Patterns of Discovery*, pp. 17, 182, n. 6 (Cambridge, 1958).

[25] See C. Kordig, "The theory-ladenness of observation", *The Review of Metaphysics*, XXIV, 3, 448–83 (Mar. 1971); *The Justification of Scientific Change*, pp. 3–13 (Dordrecht, Holland, 1971). I cannot in good conscience recommend the last work without mentioning its unfortunate repetition of one of the most serious and widespread misunderstandings concerning the history of early astronomy—the view that Copernicus simplified Ptolemaic astronomy by reducing the number of epicycles "from 84 to about 30" (p. 109). This is totally wrong, as I explain in the essay referred to in n. 14.

[26] James J. Gibson, *The Perception of the Visual World*, p. 126 (Boston, 1950).

in violent motion) the horizon may seem to be approaching or receding from the Sun; thus, for example, a character in one of Harold Pinter's plays says: "I turn. I turn. I wheel. I glide. I wheel. In stunning light. The horizon moves from the sun. I am crushed by the light."[27]

Whatever may be the superficial plausibility of asserting along with many contemporary philosophers that scientific observation is always "theory-laden", I am pleased to report that this thesis has not yet carried the field among the philosophically uninitiated. A case in point is the evident adherence to the more sober and, for me, more correct epistemology of Galileo in the refrain of John Lennon and Paul McCartney's song "The Fool on the Hill":

> But the fool on the hill sees the sun going down
> And the eyes in his head see the world spinning round.[28]

The fool on the hill is a wise fool, speaking in a thousand voices "perfectly loud" but heard by nobody; and the principal example of the fool's wisdom about which the song tells us is in the refrain: though the fool sees the sun going down, the eyes in his head (that is, the eyes of reason) see (that is, know) that it is really the world (or Earth) which is spinning round. And on this musical note I should like to close. Though I would go on, if I had time, to discuss the significance of relativity theory for the hypothesis of the Earth's motion, my conclusion would be the same: the tradition of Copernicanism—in the sense of Copernicus himself, in the sense of Kepler, Galileo, and Newton—is today still alive and healthy.[29] All honor to the memory of these illustrious men and especially, on this occasion, to the memory of the great founder of that tradition, Nicolaus Copernicus.

[27] Ellen, in Pinter's *Silence*, p. 40 (New York, 1970).

[28] The lyrics may be found on the sleeve of the Beatles' recording "Magical Mystery Tour" (Capitol Records), or in *The Beatles. Illustrated Lyrics*, ed. A. Aldridge (New York, 1969).

[29] It is sometimes alleged that Einstein's general relativity theory has the consequence that Ptolemaic and Copernican (or better, heliokinetic and geokinetic) astronomy become equally valid. This is certainly not true in any simple and unqualified sense, and indeed it may be plausibly argued that circular motion remains absolute in general relativity theory and hence that the Earth is "really" in motion. Cf. H. Bondi, *Assumption and Myth in Physical Theory*, p. 20 (Cambridge, 1967): "Since the days of Galileo and Newton we have regarded velocity as relative but acceleration as absolute. This is difficult to understand, but there doesn't seem to be any simple way out."

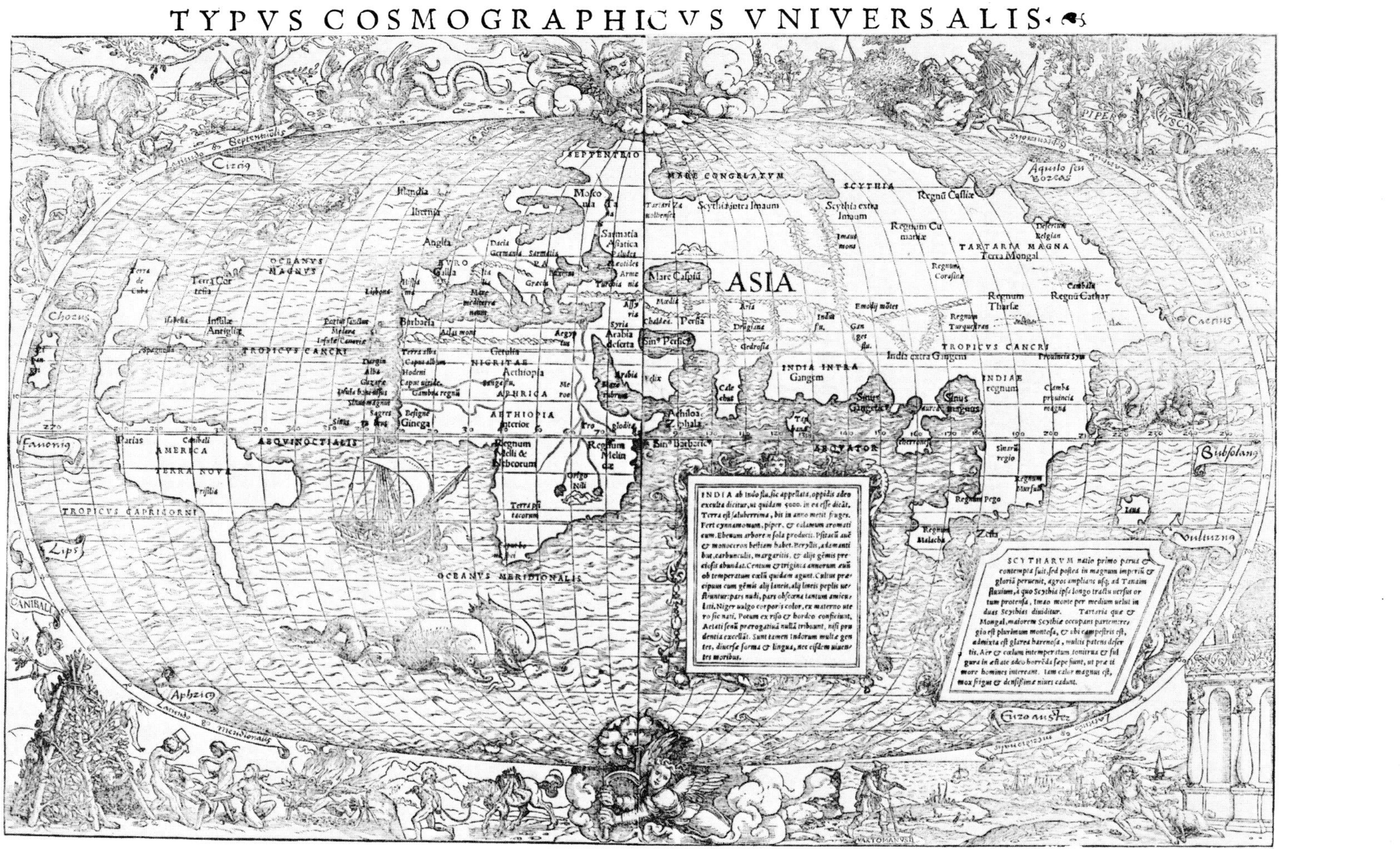

FIG. 44. Map of the world, possibly by Holbein, published in Grynaeus (or Huttich), *Novus Orbis* (Basle, 1532). (Reproduced with permission of the Humanities Research Center, University of Texas at Austin. Courtesy of Dr. R. Palter.)

Organum Uranicum.

SEBASTIANVS MVNSTERVS.

HABES IN HOC LIBRO,

AMICE LECTOR, EXPLICATAS THEORICAS OMNIVM planetarū, atq; eorundē uarios, singulos & quotidianos ad annos usq; C & ultrà expressos motus, Lunæ quoq; in lumine crescentis, senescētis & per eclipsim deficientis, & item Solis deliquiū patientis, omnē uarietatē: quibus omnibus cōmodi adiecti sunt canones.

BASILEAE APVD HENRICVM PETRVM, MENSE MARTIO, ANNO M. D. XXXVI.

FIG. 45. Title-page of Sebastian Münster, *Organum Uranicum* (Basle, 1536). (Reproduced with permission of the Humanities Research Center, University of Texas at Austin.)

SESSION 3

The Nature of Scientific Revolutions

Geocentric to Heliocentric to Galactocentric to Acentric: The Continuing Assault to the Egocentric

RICHARD BERENDZEN

Boston University

1. Geocentric to Heliocentric

Astronomy has had many great leaders, such as Tycho Brahe, the Herschels, the Struves, Henry Norris Russell, George Ellery Hale; but only one has had his name linked so frequently with the term "revolution"—Nicolaus Copernicus.[1] Although the plaudit to him has sometimes involved hyperbole,[2] it is nevertheless convenient and not too great a distortion of history to think of him as actually being "the founder of modern astronomy".[3]

This paper gives an initial consideration to whether or not the Copernican Revolution provided an archetypical model of scientific revolutions for two key developments in early twentieth-century astronomy.[4] Recently each of them has been

[1] For example, Kuhn (1957).

[2] For instance, Knight (1967), Krieger (1955), and numerous selections in Mizwa (1945).

[3] This phrase is used in the influential writings of Armitage (1957), Rosen (1971), and many others; however, a few scholars, such as Pannekoek (1948), have argued that despite his greatness, Copernicus is better thought of as closing the past rather than opening the present. Toulmin and Goodfield (1961) stress this point.

[4] More extensive information on these topics, including an analysis of pertinent archival materials, is being published by the author of this paper and his colleagues, principally, but not

called Copernican in significance. Since the details of Copernicus—his life, personality, philosophical constructs, scientific motivations, and contributions—have been extensively debated in the literature,[5] they need not be reviewed here. But at least one lesson must be drawn from these analyses: After almost 400 years of discussion and debate, unanimous agreement still has not been reached about Copernicus' contributions to the development of science. Surely twentieth-century astronomy, too, will require lengthy debate before it can be viewed in proper perspective; nevertheless, even from the myoptic range of 50 years, many lessons can be learned. The time for assessment is already at hand.

2. Heliocentric to Galactocentric

2.1. *The Universe, ca. 1900*

In 1890, Agnes Clerke, one of the most informed and respected writers on astronomy of the time, gave the following summary of the prevailing opinion about the universe's large-scale structure.

> The question whether nebulae are external galaxies hardly any longer needs discussion. It has been answered by the progress of discovery. No competent thinker, with the whole of the available evidence before him can now, it is safe to say, maintain any single nebula to be a star system of coordinate rank with the Milky Way. A practical certainty has been attained that the entire contents, stellar and nebular, of the sphere belong to one mighty aggregation, and stand in ordered mutual relations within the limits of one all-embracing scheme.[6]

At the beginning of the twentieth-century, almost all astronomers agreed that the Sun was located near, but not precisely at, the center of the Milky Way, and that the Milky Way constituted the entirety of the universe.[7] The generally accepted model[8] for the system was that of William Herschel,[9] developed a century earlier (Fig. 46, p. 82); it had resulted from painstaking observations and calculations, using a straightforward but naïve concept of stargauges. The Herschelian Universe placed

---

exclusively, in the *Journal for the History of Astronomy*, vol. 6, 1974. These papers are related to the Case Studies Project on the Development of Modern Astronomy (Berendzen, 1972, p. 146).

[5] For instance, in Birkenmajer (1900, 1924, 1966), Burtt (1955), Butterfield (1959), Cohen (1960), Dreyer (1953), Frank (1944), Hall (1970), Koestler (1963), Koyré (1942), Kuhn (1957, 1970), Prowe (1883), Rosen (1971), and many other sources. Of these, the standard, nearly definitive biography is the one by Prowe (1883). An extensive annotated bibliography, especially for 1939–58, on materials pertaining to Copernicus is given in Rosen (1959).

[6] Clerke (1890).

[7] See North (1965, chap. 1), and Struve and Zebergs (1962, chaps. 19, 20).

[8] For simplicity the term "model" is used predominantly in this paper. A more detailed analysis of the scientific revolutions would necessitate deliberation of when "model" should be used in contrast with "system", "plan", "theory", "hypothesis", and other terms. But such a discussion is not pertinent to this paper.

[9] See Hoskin (1963).

the Sun close to the center of a flattened "stratum of the fixed stars", which presumably was five times larger in breadth than in thickness.

Starting in about 1901, J. C. Kapteyn, and later his associate P. J. van Rhijn, revived and improved Herschel's method, culminating in the Kapteyn Universe.[10] This model, although based on a more elaborate and sophisticated technique than Herschel's, led to a universe of essentially the same structure (Fig. 47, see p. 82).

An independent but related question dealt with the nature of spiral nebulae. Since the seventeenth-century, astronomers and philosophers alike had speculated about these faint nebulous objects. As early as the eighteenth-century some philosophers had suggested that there may exist numerous systems comparable with the Milky Way in size and structure—"island universes". But, scientific arguments were marshalled against such a possibility; even so, the concept lingered for many decades, and by the early part of the twentieth century new evidence seemed to support the old concept.

The basic difficulty in understanding spirals, as well as the Milky Way, was distance determination. In 1900, no one knew for certain whether spirals were small nearby objects or large remote ones. If the former case were true, the spirals then would lie within the Milky Way, being unusual but not exceptional members of the general stellar system, all of which presumably was centered near the Sun; but if the latter case were true, the spirals then would lie outside the Milky Way, becoming "island universes".

These two issues—the location of the Sun and nature of spirals—caused controversy during the late teens, culminating in 1920 in astronomy's so-called Great Debate.[11]

2.2. *Astronomy's Great Debate*[12]

(a) THE SUN'S LOCATION

Surely, since antiquity the Milky Way has been one of the most impressive sights in the clear night sky; and since 1609 astronomers have had direct evidence that it is composed of numerous individual stars. But its size, shape, and structure, as well as whether or not it is unique, eluded astronomers until about 50 years ago. One of the issues involved in resolving these problems was the place of the Sun in the system.

In 1917, on the basis of a study of globular clusters and with the newly devised period–luminosity curve for Cepheids, Harlow Shapley, a gregarious young American

[10] Kapteyn and Rhijn (1922).

[11] Otto Struve, among others, has applied this term; see Struve and Zebergs (1962, p. 416).

[12] A highly edited transcript of the debate was published a year later: Shapley and Curtis (1921). A detailed study of the debate and the scientific events surrounding it will be published by Berendzen and Hoskin in 1975, in 2 parts, in the *Journal for History of Astronomy*, volume 6.

astronomer at the Mt. Wilson Observatory, made the first serious challenge to the Herschel–Kapteyn models. He boldly proposed that our Galaxy is a lens-shaped system, some ten times larger in scale than anyone had previously believed, with the Sun located roughly 50,000 light years from the center. This new model was known as the "galactocentric" universe (Fig. 48 on p. 83).

Peter van de Kamp, Chairman of the Astronomy Department at Swarthmore College, has recently written that "The galactocentric viewpoint developed by Harlow Shapley in 1917 represents a step forward in astronomical thinking and perception analogous to the introduction of the heliocentric viewpoint by Nicolaus Copernicus in 1543."[13]

(b) NATURE OF SPIRALS

But Shapley's proposals had not gone unchallenged. His principal opponent was H. D. Curtis, a believer in the Kapteyn model. Moreover, from his extensive study of spirals, Curtis concluded that they must be remote. Shapley, on the other hand, believed that the spirals are relatively small and nearby, probably within the Milky Way system. But the early arguments of Curtis, later supported by key observations of Edwin Hubble, proved the spirals to be external galactic systems.

George Abell, Chairman of the Astronomy Department at U.C.L.A., recently wrote in his *Exploration of the Universe*, that "The realization, only a few decades ago, that our Galaxy is not unique and central in the universe ranks with the acceptance of the Copernican system as one of the great advances in cosmological thought."[14]

(c) CONFLICTS

Thus, in 1920, two individuals—Curtis and Shapley—symbolized the opposing camps on two related but separable issues. The achievements on each of these issues have now been likened to the Copernican Revolution; hence, in 1920, the world saw not just one but two allegedly Copernican-style controversies!

To make the matter even more dramatic, the two protagonists actually met in a face-to-face debate, held on April 26, 1920 at a meeting of the National Academy of Sciences in Washington, D.C.[12]

It is ironic, or may be it only seems so today, but the debate almost did not take place.[15] In planning for the annual meeting of the Academy, George Ellery Hale

[13] P. van de Kamp (1965).

[14] G. Abell (1969). This is probably the largest-selling astronomy book ever written.

[15] The specifics noted here are disclosed in unpublished archival sources, which will be included in the forthcoming article in footnote 12.

had proposed that there be some type of debate, since such events usually are stimulating to audiences. One of the topics considered was relativity, but in January the arrangers of the meeting decided, perhaps only partially in jest, that the controversy about relativity might well be settled before the meeting took place in April. They finally decided that the question of island universes, or possibly the scale of the universe, would make a suitable topic. It is also a curious historical fact that the persons originally proposed were Shapley and W. W. Campbell, who was then Director of the Lick Observatory where Curtis was employed.

Eventually the program was settled, with Curtis and Shapley as the participants. Although today the event is frequently called a "debate", at the time it was formally listed as a "discussion". Certainly it was not a debate in the usual sense.[16] Each of the participants delivered his paper (with Curtis largely reading his from typed slides), and their formal statements were followed by brief rejoiners.

Even though the view is imperfect, the mental imagery is arresting to picture a modern-day Copernicus and Ptolemy squaring off in direct personal confrontation. It is like Galileo's *Dialogue* come alive!

Since there were two topics, each now termed Copernican in significance, one might reasonably suppose that this event would have attracted enormous attention, probably of the public but certainly of scientists. Remarkably, that was not the case. Although articles about it were carried in a few key newspapers (such as the *Boston Globe*, *London Times*, and *New York Times*), the event roused scant attention; indeed, it was not until over a generation later, when Otto Struve and others began appraising the event with some historical perspective, that its significance was realized.[17]

3. Galactocentric to Acentric

3.1. *Resolution of the Conflict*

Like most debates, this one did not settle the disagreements; rather, it brought a few matters to focus and more clearly identified the problems. In the years immediately following the Debate, astronomers remained divided.

The definitive solution about the nature of spirals required new data. They were not long in coming. As Sandage has succinctly (but oversimplistically) put it: "What

[16] See Berendzen (1970). Hubble (1936) referred to it as a "quasi-debate", but he did not explain the term.

[17] In fact, even the participants themselves failed to comprehend the moment of the occasion, as is made clear by a careful reading of their private letters from that era. Shapley (1969, p. 79) emphasized this in later reminiscences. The significance of an event frequently is difficult to realize while it is transpiring; e.g. on December 30, 1972, at the meeting of the American Association for the Advancement of Science in Washington, D.C., a discussion of potentially enormous import occurred—conceivably rivalling the Great Debate—when Halton Arp and John Bahcall debated the correctness of using red shifts as distance indicators in cosmology.

are galaxies? No one knew before 1900. Very few people knew in 1920. All astronomers knew after 1924."[18]

In late 1923 and during 1924, Hubble obtained photographs of Cepheids in spirals; then, employing Shapley's period–luminosity relationship, he calculated their distances and found them to be enormous. During 1924 his preliminary findings were discussed among most of the galactic astronomers in the United States, and as early as November of that year the *New York Times*[19] carried an article about his findings. But the formal presentation was made in a paper read on his behalf on 1 January, 1925 at a joint meeting in Washington, D.C. of the American Astronomical Society and the American Association for the Advancement of Science.

Although Hubble's dramatic findings supplied incontrovertible evidence for the extragalactic nature of spirals, the calibration of the crucial period–luminosity curve and the measurement of the size and structure of spirals, as well as of our own galaxy, were yet to be achieved with precision. But at least in gross terms, 1920 marked the epicenter of the conflict, while 1925, only five years later, yielded the resolution. Thus by the mid-1920s, the galactocentric model was established and the remoteness of spirals proved; concomitantly, with the largest changes since Renaissance astronomy, man's Earth had become even more peripheral and miniscule.[20]

3.2. *The Expanding Universe*

But then a new question arose: If spirals are external systems comparable with the Milky Way galaxy, what is the larger structure of the Universe, i.e. how does the local system fit within the metagalaxy? A complete answer has yet to be found, but the beginnings of one came before 1930.

Starting in the mid-teens, V. M. Slipher at the Lowell Observatory had obtained spectra of spirals, on which he discovered that their velocities were preferentially recessional at extraordinarily high speeds. For about 10 years, he was virtually the only astronomer working on the radial motions of galaxies. In the late 1920s, however, Hubble (with his assistant, M. Humason) expanded Slipher's observational data, and coupled measured motions of spirals with his own calculated distances for them, which enabled him to announce in 1929 one of the most dramatic findings of twentieth-century astronomy—that there is an approximately linear relation between the radial velocity of extragalactic nebulae and their distances.[21] Even though Hubble's

[18] Sandage (1961, p. 1). An examination of pertinent archival sources is given in Berendzen and Hoskin (1971).

[19] *Times* (1923).

[20] Shapley (1958, 1963, 1969) has discussed these changes in several books and articles. In keeping with his fondness for polemics, he often spoke with delight about his contribution to attacking man's "egocentricism". It is interesting to compare Shapley's remarks about his own achievement with Blumenberg's (1955) analysis of the Copernican shift.

[21] Hubble (1929).

values for galactic distances later proved to be in serious error, he had already begun to establish the cornerstone of modern cosmology, viz. that the universe is expanding.

In contemporary times probably no other finding has so greatly influenced astronomy; and its ramifications have touched physics, mathematics, theology, and other fields. Within astronomy itself, it has provoked extensive discussion of cosmological and cosmogonical models; like Neo-platonists, the modern theorists strive to "save the phenomena". Although the controversies still rage, one generally accepted conclusion from Hubble's early work is that the expanding Universe is, in fact, acentric. The cosmos now appears to be enormous but finite in both extent and duration, yet man's galaxy is at no privileged point in either space or time (Fig. 49, see on p. 83).

4. Brief Comparison of these Scientific Revolutions

In this section the nature of the Copernican Revolution is briefly compared with those of the galactocentric and acentric revolutions. To do so requires delineating a few characteristics by which any scientific revolution might be considered. The ones discussed here are typical of those that have been mentioned in the literature,[22] explicitly or implicitly.

4.1. *Did the Revolution Occur at a Time of Scientific Crisis?*

(a) COPERNICAN

Textbooks occasionally simplify the story, suggesting that Copernicus devised his model at a time of great scientific conflict, when the Ptolemaic scheme was increasingly being found in error. Despite its appeal, this explanation is not adequately complex, as Gingerich[23] and others[24] have noted. Certainly, it was known by Copernicus' time that the Ptolemaic model was not flawless, being incapable of making totally reliable long-term ephemerides. But astronomy was not in a state of crisis; confusion, perhaps, but crisis, no.

(b) GALACTOCENTRIC AND ACENTRIC

At the beginning of the twentieth century there was no major controversy regarding the place of the Sun or the nature of spirals. This tranquil atmosphere changed

[22] For instance, see Barber (1963), Birkenmajer (1966), Carlson (1972), Frank (1944), Hall (1963), Kuhn (1970, especially pp. 174 ff), Purtill (1967), Shapere (1964), and Toulmin (1967).

[23] Gingerich (1972).

[24] Birkenmajer (1966), Burtt (1955, chap. II). It should be noted, however, that growing pressure for calendar reform in Copernicus' time led to a certain amount of crisis; see Kuhn (1957, chap. 4).

by the late teens. By then, confusing, conflicting information had been accumulated on both of these issues; consequently, by 1920 galactic astronomy had reached a state of mild controversy. A careful examination[4] of virtually all extant records, plus interviews today with senior astronomers from that period, indicate that most of the world's astronomers were interested in, but not ardently aroused or completely convinced by, any of the arguments. Thus the revolution of Curtis, Hubble, and Shapley came at a time of astronomical controversy and conflict, but not crisis.[25]

4.2. *Was the Revolution Prompted by or Based on New Data?*

(a) COPERNICAN

The Copernican system did not rest upon new observational data.[26] It was not formulated to fit new findings, but rather for philosophic and aesthetic reasons.

(b) GALACTOCENTRIC AND ACENTRIC

These twentieth-century revolutions depended fundamentally upon new observational data, as well as upon a new empirical technique, the period–luminosity law. Among the mandatory data were the observations of Cepheids in globular clusters and in spirals, the detection of novae in spirals, and the measurement of Doppler shifts of spirals.

4.3. *Did the Revolution Occur because of Improved Technology?*

(a) COPERNICAN

To decide between the Ptolemaic and Copernican models, on the basis of accuracy of prediction, would have required more precise data than existed at the time of Copernicus, or for decades afterwards. Adequately precise data eventually were provided by Tycho, whose success in this regard came not only because of his relatively superior sighting devices at Uraniborg, but also because of his meticulous care in observing and assiduousness in providing error bounds.

[25] Even though this conclusion follows from a thorough modern study of the entire astronomical community in *ca.* 1920, it is perplexing to find that a few weeks before the actual debate, Shapley (1920) asserted in a letter to H. N. Russell that E. W. Brown, E. B. Frost, and several other astronomers viewed the "coming discussion as a *crisis* for the newer astrophysical theories" (emphasis added). Lest too much be inferred from this, however, it should be noted that Shapley's letters of that time frequently overestimated the impact that his findings were having. Neither the private nor the public papers of other astonomers at that period fully corroborate his zealous declarations of a sense of revolution and crisis, resulting from his iconoclasm.

[26] Sarton (1953) has put it thus: "Copernicus was a poor observer and it had been easier for him . . . to formulate his new theory, because he was not embarrassed by good observations."

(b) GALACTOCENTRIC AND ACENTRIC

The breakthroughs of Curtis and Shapley were made at the new mountain observatories in California—Lick and Mt. Wilson.[27] Each of these men used some of the finest astronomical instruments then available in the world, but elementary calculations show that their findings probably could have been achieved with much more modest instruments. Likewise, Hubble's greatest observational findings in the 1920s, made with the world's two largest telescopes, the 60- and 100-inch reflectors, could have been achieved with smaller instruments.

But in fairness to the other observers it must be noted that the California observatories excelled in many respects, size of telescopes being but one. The optics of those instruments were of outstanding quality, and they were maintained by talented professional staffs; plus, the seeing there was unusually good. Thus, even though the discoveries were not entirely aperture-bound, these large, first-rate instruments greatly aided the work.[28]

Also, these breakthroughs depended crucially upon the nineteenth-century technological advances of photography and spectroscopy.

[27] A discussion of the development of these mountain observatories has been given by A. E. Whitford in Berendzen (1972, p. 202).

[28] The issues involved here are not only complex but also moot. Wright, Warnow, and Weiner (1972) have matter-of-factly stated that: "without the 100-inch telescope [Hubble's discovery of Cepheids in spirals] would have been impossible at that time." In his own discussions of the matter, however, Hubble (1926, 1936) did not make that claim. Instead, he reviewed the physical problems involved, which clearly required a large, sensitive telescope, but not necessarily the 100-inch. To achieve adequate light-gathering power, an aperture of only 20 inches or so would have been adequate. By the early 1920s several telescopes of this size or larger were operative. But a more severe problem lay in resolving individual faint images in spirals, for at that time no one knew for certain whether they were normal stars in remote systems or nebulous patches in relatively nearby, amorphous objects. Thus, in principle, the improved resolving-power offered by the 100-inch made it indispensable. But it is doubtful that any of the instruments would have been diffraction-limited; instead, resolution probably was set by the seeing conditions and photographic plates. Even so, the 100-inch was still probably helpful, because its locale had superb seeing and its fast mirror permitted comparatively short exposures on fine grain plates. Obviously, therefore, the 100-inch telescope, coupled with its fine site and ancillary equipment, must have greatly aided Hubble; nevertheless, theoretically at least, a smaller instrument at a less propitious location should have been adequate. For instance, C. D. Shane (1973), the former director of the Lick Observatory, recalls that, "when Hubble visited Mr. Hamilton, probably in January 1947, he examined a plate of the Galaxy recently taken with the 20-inch astrograph and pointed out the first Cepheid he identified in it".

In the late 1920s, Hubble began to press observational cosmology to the limit of the 100-inch. Unquestionably, by then no other instrument could peer so deeply into space, revealing the velocity–distance relationship. But as the search went to even fainter objects, ultimately even the 100-inch became inadequate. The yearning to extend the boundary of the observable universe ever farther became one of the prime arguments for constructing the 200-inch reflector.

4.4. *Did the Revolution Occur because of New Findings in Science or Mathematics?*

(a) COPERNICAN

Copernicus' system, regardless of its innovativeness,[29] did not involve new fundamental concepts.[30] As Bertrand Russell[31] has put it, "there is nothing in his speculations that could not have occurred to a Greek astronomer". Obvious examples of new understandings that were to come later would be Kepler's elliptical motions and Newton's gravity.

(b) GALACTOCENTRIC AND ACENTRIC

Except for the period–luminosity law, which depends upon astrophysical theory that was not understood at the time, these revolutions were primarily observational.[32]

4.5. *Did the Revolution Require Boldness to Confront Peer Pressure?*

(a) COPERNICAN

The assertiveness that Copernicus must have needed is still a moot point, but at least one conclusion seems safe: By Copernicus' time neither Ptolemy nor Aristotle was sacrosanct; nevertheless, there were numerous, well-known arguments against the heliocentric model, which Copernicus could answer scarcely better than had Aristarchus centuries earlier.[33]

(b) GALACTOCENTRIC AND ACENTRIC

Shapley had to possess considerable nerve to confront the astronomical establishment of his time, for by *ca.* 1920 the Kapteyn model was widely accepted, while Shapley's proposal was truly radical. It depended upon a vaguely understood empirical relationship and employed dubious calibration techniques, of only eleven stars; moreover, his entire model was based on a total of sixty-nine globular clusters. A less audacious, ebullient person than the young Shapley likely would have been more cautious; in fact some of his senior colleagues, such as Russell, cautioned him against intrepid impetuosity.[4]

[29] See Kuhn (1957, chap. 5).
[30] See, for example, Butterfield (1959, chap. 2).
[31] Russell (1959, p. 526).
[32] The theoretical work of Eddington, Jeans, and a few others was pertinent but marginally influential, and certainly not critical.
[33] Burtt (1955, pp. 36 ff.).

Curtis was a relative conservative. His views about the Milky Way were in keeping with those of Kapteyn and most other astronomers, and his arguments about spirals did not encounter major conflict with any astronomers other than Shapley and his colleague A. van Maanen.

Likewise, Hubble's findings until the late 1920s were epochal but not disturbing. But his expanding-universe hypothesis became extremely controversial, even bringing him into conflict with Einstein. In fact, Hubble himself vacillated in his beliefs about the hypothesis, until more findings during the 1930s established it with certainty.

4.6. *Did the Scientific Revolution Mesh with the Existing Social Ethos?*

(a) COPERNICAN

Even though heliocentrism and its corollaries were contrary to contemporary social beliefs, "(the Church) took nearly fifty years to see that Copernicus was bound to lead to Voltaire".[34] And, in fact, the main conflict did not arise until 50 years after that; once the clash occurred, however, it was severe, becoming a notorious episode in Western intellectual history. Nevertheless, in the years immediately following Copernicus, his model did not cause appreciable societal conflict.

(b) GALACTOCENTRIC AND ACENTRIC

Interestingly, neither the galactocentric nor the acentric models seemed to have aroused a response from society, a situation in marked contrast with that for Darwinism and Freudianism. The popular press and editorial writers almost totally ignored the galactocentric model, and although the acentric concept and the expanding universe stirred popular attention, they did not promote outcries from theologians, politicians, editorial writers, or other spokesmen for the public.

4.7. *Did the Revolution Simplify Scientific Models?*

(a) COPERNICAN

Even though one of Copernicus' prime motivations was a desire for simplicity, in order to make his model work without abandoning circular motion, he arrived at a heliocentric model as complex as the archaic Ptolemaic one, which he had hoped to replace.[35]

[34] Butterfield (1959, p. 56).

[35] No one has shown this more lucidly and tersely than Cohen (1960, p. 58). It has also been proved in detail by Neugebauer (1968), following the classic works of Kuhn (1957) and of Price (1959). The older, conventionalistic account was frequently given before 1960; e.g. by Polanyi (1951).

(b) GALACTOCENTRIC AND ACENTRIC

With the heliocentric and acentric systems, many observational findings began to fit into a logical whole. For instance, in the nineteenth century, spectroscopic studies of nebulae had proved confusing, showing some with continuous spectra and others with lined spectra. Curtis' determination of the true nature of spirals helped to make these observations understandable.

4.8. *How Rapidly Was the Revolutionary Model Accepted?*

(a) COPERNICAN

Scientific theories can be accepted by two different groups: scientists and the public. For the Copernican model, several decades passed before even a few bold astronomers embraced it, and almost 150 years lapsed before most scientists were fully convinced.[36] And the public's reaction was similar, but the time span even greater. An obvious cause for this was the organized opposition by the Church and other institutions.

Acceptance of any scientific theory can be roughly gauged by its treatment in textbooks. Boutroux[37] has shown that in France, at least, the textbook writers were extremely cautious, some of them rejecting Copernicanism as late as *ca.* 1700.

(b) GALACTOCENTRIC AND ACENTRIC

As mentioned in Section 3.1, astronomers accepted the galactocentric model plus the island universe concept as soon as observational data established them beyond reasonable doubt, which happened to occur only a few years after the concepts were seriously debated. The expanding universe, however, was a slightly different case. It did not arise from philosophy, theory or indirect reasoning, but was based on observable fact.[38] Consequently the observations were readily believed, but the interpretations of them are still being debated.

The public reaction, however, was slightly different. Examination[4] of the leading newspapers and magazines in the United States and England from 1900 to 1940 reveals that the galactocentric galaxy caused almost no discussion whatsoever, while the expanding Universe prompted numerous articles. Likewise, the author of this

[36] Butterfield (1959, p. 55).

[37] Boutroux (1922). See also Kuhn (1957, pp. 188 ff.). An informative but not entirely reliable account of resistance to Copernicanism has been given by Scott (1954).

[38] What was directly observable was that the sizes of the red shifts of spirals correlated with their distances. Since most astronomers (then as well as now) believe these red shifts to indicate speed in a straightforward Doppler fashion, the observations directly indicate an expanding universe. But if the shifts are non-Doppler, other possibilities arise.

paper has investigated every astronomy textbook in the U.S. Library of Congress to measure the influence of these modern revolutions. The galactocentric one became fairly well treated in astronomy texts within a decade of the Great Debate, although the Debate itself was rarely mentioned. The expanding universe was picked up even more promptly by textbook authors, albeit sometimes in confused ways. As late as the 1940s, the writers frequently would print errors about these events in the 1920s; but certainly they did present those developments as being scientifically worthwhile.

Why, then, was the galactocentric model almost completely ignored by the public? Several tentative answers can be proffered:

(i) The vast distances and peculiar geometry of this model are conceptually non-trivial.[39] This point should be significant to both educators and historians of science, because recent studies[40] have proved that non-scientists frequently have lower abilities at three-dimensional conceptualization than do scientists.

(ii) By the mid-1920s, the world had witnessed a constellation of scientific breakthroughs, some bizarrely defying logic: those of Bohr, Darwin, Einstein, Freud, Planck, to name but a few.[41] The galactocentric shift, by comparison, must have seemed a relatively mundane transformation. Indeed, the Copernican Revolution itself was the intellectual ancestor of these desensitizers.

(iii) Kuhn has ventured that "After a science has become thoroughly technical, its role as a force in intellectual history becomes relatively insignificant."[42] Although the galactocentric model was not so technical as some of the contemporary theories in physics, it was complex enough to lessen its public impact.

(iv) Long before the twentieth century, man already had been thoroughly convinced of his physical smallness.[43] In the nineteenth century, for instance, even before Shapley had expanded the scale, Thomas Carlyle summarized this feeling about the Universe: "It's a great deal bigger than I am. . . . People ought to be modester." In terms of characteristic mass, size, and time, man lies roughly midway

[39] Struve (1942) has argued that the galactocentric model required a rejection of common-sense reasoning similar to Copernicus' rejection of it.

[40] Siemankowsky and MacKnight (1971).

[41] Timing is of the essence; being either early or late can prove calamitous. As Badash (1966) has demonstrated about the reception in 1902 of the Rutherford–Soddy theory of radioactive transmutations, if its announcement had come earlier, the acceptance would have been less. So, too, for the galactocentric model. Before the period–luminosity law, it would have been strictly metaphysical; after Hubble in 1924, it would have been entirely anticlimatic.

[42] Kuhn (1971).

[43] A recent exchange of letters in *Science* may suggest, however, that this conclusion is not entirely acceptable. To Cole's (1972) claim that ". . . the self-esteem of the average person has survived intact the revelation that the earth is not the center of the universe. It takes a rare person to spend more than a few hours pondering the philosophical implications of that proposition . . ." a secretary (Holmes, 1973), who characterized herself as being a "very 'average' person", replied that ". . . there is no way that I could account for all the hours I wondered and worried and tried to understand our fragile position here on earth. As for the people who are not vocal about it—believe me, they feel it."

between the atomic nuclei and galaxies. But with (perhaps even because of) his growing awareness of this scheme of things, he also began to comprehend the power of his most outstanding attribute—the ability to think. As man grew smaller, his mind grew larger.[44] Thus, ironically, the galactocentric model proved to be a less severe blow to egocentricism than Shapley might have thought or hoped for.

(v) The galactocentric model did not pose a direct threat to any societal institution, as the Copernican had to seventeenth-century religion.[45] It is clear from the widespread reactions to Copernicanism, Darwinism, and Freudianism that when scientific revolutions impinge upon metaphysics or social theory, they are likely to become unusually polemical and possibly unacceptable.

The expanding Universe, in contrast, was fundamentally more exciting and intriguing than an eccentric sun; moreover, it had philosophical and theological implications, and it was closely allied to other scientific revolutions, such as relativity theory.

5. Conclusions

Was the Copernican Revolution archetypical for these twentieth-century revolutions? Even to begin to answer that requires first that two other questions be answered: What is the nature of scientific revolutions in general? What was special about the Copernican Revolution?

The characteristics discussed in Section 4 are typical diagnostics for understanding scientific revolutions in general. Clearly, no definitive criteria, yielding necessary and sufficient characteristics for all such revolutions, can be devised; human enterprises are never that programmable. But analysis seems to support the thesis that the breakthroughs of Curtis, Hubble, and Shapley could correctly be termed revolutionary.

But how Copernican were they? Dingle has eloquently posited that what Copernicus undertook was nothing less than "the complete reform of astronomy; the erasure of the product of two thousand years of toil, and a rebuilding of the whole edifice. Such a task had never been attempted before; it had not needed to be attempted since."[46] Even after slicing away the exaggeration of this assertion, some truth to it lingers. Certainly these twentieth-century developments were of no comparable moment. In that and other respects they were non-Copernican, as Section 4 has indicated. One of them was ignored by the public, while the other stirred only

[44] The few items in the *N.Y. Times* from 1917 to 1940 that indicated public interest in the galactocentric revolution and expanded distance scale expressed views similar to this one. See *Times* (1930, 1938).

[45] Struve (1942) claimed that the galatocentric concept was accepted more readily than the Copernican had been because by the twentieth century, unquestioning devotion to authority had been expunged from science and the Catholic Church's influence had waned.

[46] Herbert Dingle in Rapport and Wright (1965, p. 22).

minor attention; these, then, cannot rightfully stand in juxtaposition with Copernicanism, which ultimately aided in re-orienting man's understanding of the world and of himself.

But Copernicus' greatest contribution "lay not so much in the system he propounded as in the fact that the system he did propound could ignite the great revolutions in physics" of Galileo, Kepler, and Newton.[47] In that respect, which is possibly the ultimate meaning of Copernicanism, these twentieth-century revolutions were Copernican, although on a smaller scale. Only the historians of the future can gauge the influence of these revolutions when weighed by the sternest arbiter of all—time.

Will there be other Copernican Revolutions? North has observed that "we have found no evidence of immortality in the natural sciences. None has an absolute and permanent value. The individual theory of cosmology is neither true nor false: like any other scientific theory, it is merely an instrument of what passes for understanding".[48] While no other scientific revolution may ever again have the total impact of the Copernican, there will indeed be a continuum of other revolutions, like those discussed here, each contributing its own increment. To believe otherwise would be to reject massive evidence;[49] but besides that, it is the quest for revolutions—for breakthroughs that permit new understandings and knowledge—that gives science life and makes it worth doing.

Acknowledgments

I should like to thank the following individuals for their helpful comments on a preliminary draft of this article: L. Badash, O. Gingerich, M. A. Hoskin, C. D.

[47] Cohen (1960, p. 63).

[48] North (1965, p. 407).

[49] In 1970 the retiring president of the American Association for the Advancement of Science (Glass, 1971a) argued that science is finite and that it is rapidly approaching completeness, when all fundamental concepts will be known. Naturally, this eristic position prompted retorts (Fink, 1971; Meade, 1971; Steinberger, 1971), as well as clarification by its progenitor (Glass, 1971b).

At least since the time of Newton, divers seers, after surveying the state of scientific progress, have soberly concluded that man was then completing the final chapter of attainable knowledge. And, as if that declaration itself was heuristic, almost immediately thereafter new scientific discoveries have shaken the accepted world view. Such predictions of completeness are now being scrutinized by historians; for instance, Badash (1972) has reviewed some of them, using nineteenth-century science as a case study.

It would indeed appear that if a Universe's physical constituents are finite, so too must be its properties and laws—that is, its science. But, in this writer's view, no persuasive case has been made that man is approaching that nirvana. In fact, perhaps virtually by definition no such proof can ever be constructed: a fully convincing argument may only be possible *a posteriori*. But even then it would doubtlessly be one of those nemeses of logic—a negative proof. On this matter, therefore, agnosticism may be the prudent course.

Another irony wed to this debate relates to the desirability of ever achieving total knowledge, assuming that it were possible. The great edifice of scientific inquiry would then be complete. Paradoxically, science seeks full understanding, yet if it were reached, science would cease.

Shane, and D. Seeley. The research for this paper was partially supported by a grant from the National Science Foundation.

References

ABELL, G. (1969) *Exploration of the Universe*, Holt, Rinehart & Winston, New York, p. 608.

ARMITAGE, A. (1957) *Copernicus: The Founder of Modern Astronomy*, Thomas Yoseloff, New York.

BADASH, L. (1906) "How the 'Newer Alchemy' was received", *Scientific American*, **215**, no. 2, 88–95.

BADASH, L. (1972) "The completeness of nineteenth-century science", *Isis*, **63**, 48–58.

BARBER, B. (1963) "The structure of scientific revolutions", *Amer. Sociological Review*, **28**, no. 1, 298.

BERENDZEN, R. (1970) Review of Shapley (1969), *J. Hist. Astronomy*, **1**, 85.

BERENDZEN, R. (ed.) (1972) "Education in and history of modern astronomy", *Ann. N.Y. Acad. Sciences*, **198**.

BERENDZEN, R. and HOSKIN, M. (1971) "Hubble's announcement of cepheids in spiral nebulae", *Leaflets of Astron. Soc. Pacific*, no. 504.

BIRKENMAJER, L. A. (1900) *Mikolaj Kopernik*, Krakow.

BIRKENMAJER, L. A. (1924) *Stromata Copernicana*, Krakow.

BIRKENMAJER, L. A. (1966) "Elementry tradycjonaline i nowatorskie w kosmologii Mikolaja Kopernika", *Kwart. Hist. Nauki, Tech. II*, 13.

BLUMENBERG, H. (1955) "Der Kopernikanische Umsturz und die Weltstellung des Menschen", *Studium Generale*, 8, 637.

BOUTROUX, P. (1922) "L'enseignement de la mécanique en France au XVIIe siècle", *Isis*, **4**, 276.

BURTT, E. A. (1955) *The Metaphysical Foundations of Modern Science*, Anchor, New York.

BUTTERFIELD, H. (1959) *The Origins of Modern Science*, Macmillan, N.Y.

CARLSON, E. (1972) Ph.D. thesis, unpublished, Boston University, Boston, chap. VI.

CLAGETT, M. (ed.) (1959) *Critical Problems in the History of Science*, Univ. of Wisconsin, Madison.

CLERKE, A. (1890) *The System of the Stars*, p. 368, Longmans, Green & Co., London.

COHEN, I. B. (1960) *The Birth of A New Physics*, Anchor, New York.

COLE, L. S. (1972) Letter, *Science*, **178**, 561.

DREYER, J. L. E. (1953) *A History of Astronomy from Thales to Kepler*, 2nd ed., chap. XIII, Dover, New York.

FINK, B. R. (1971) Letter, *Science*, **172**, 111.

FRANK, P. (1944) "The philosophical meaning of the Copernican revolution", *Proc. American Philosophical Soc.* **87**, 381.

GINGERICH, O. (1972) "Demythologizing the Copernican Revolution", talk at Boston University, October 18; and " 'Crisis' versus aesthetic in the Copernican revolution", this volume, p. 85.

GLASS, B. (1971a) "Science: endless horizons or golden age?", *Science*, **171**, 23–29.

GLASS, B. (1971b) Letter, *Science*, **172**, 111–12.

HALL, M. B. (1963) "The structure of scientific revolutions", *The American Review*, LXVIII, no. 3, 700.

HALL, R. J. (1970) "Kuhn and the Copernican revolution", *Brit. J. Phil. Sci.* **21**, 196.

HOLMES, A. E. (1973) Letter, *Science*, **179**, 329.

HOSKIN, M. A. (1963) *William Herschel and the Construction of the Heavens*, Norton, New York.

HUBBLE, E. (1926) "A spiral nebula as a stellar system, M33", *Astrophys. J.* **63**, 236–74.

HUBBLE, E. (1929) "A relation between distance and radial velocity among extra-galactic nebulae", *Proc. Nat. Acad. Sci.* **15**, 168.

HUBBLE, E. (1936) *Realm of the Nebulae*, p. 88, Yale Univ., New Haven.

KAMP, P. VAN DE (1965) "The Galactocentric Revolution", *Publ. Astron. Soc. Pacific*, **77**, 325.

KAPTEYN, J. C. and RHIJN, P. J. VAN (1922) "First attempt at a theory of arrangement and motion of the siderial system", *Astrophys. J.*, LV, 302.

KNIGHT, D. (1967) *Copernicus: Titan of Modern Astronomy*, Chatto & Windus, London.

KOESTLER, A. (1963) *The Sleepwalkers*, Part 3, Grosset & Dunlop, New York.
KOYRÉ, A. (1942) "Nicholas Copernicus", *Bull. Polish Inst. Arts and Sciences in America*, **1**, 705.
KRIEGER, E. (1955) "Nikolaus Kopernikus—Begründer unseres Weltbildes", *Ostdeutsche Monatshefte*, **22**, 13.
KUHN, T. S. (1957) *The Copernican Revolution*, Vintage, New York.
KUHN, T. S. (1970) *The Structure of Scientific Revolutions*, 2nd ed., Univ. of Chicago, Chicago.
KUHN, T. S. (1971) "History and history of science", *Daedalus*, Spring issue, p. 277.
MEADE, G. P. (1971) Letter, *Science*, **172**, 111.
MIZWA, S. P. (ed.) (1945) *Nicholas Copernicus, A Tribute of Nations*, Kosciuszko Found., New York.
NEUGEBAUER, O. (1968) "On the planetary theory of Copernicus", *Vistas in Astronomy*, ed. A. Beer, **10**, 89.
NORTH, J. D. (1965) *The Measure of the Universe*, Clarendon Press, Oxford.
PANNEKOEK, A. (1948) "The planetary theory of Copernicus", *Popular Astronomy*, **56**, 2.
POLANYI, M. (1951) *The Logic of Liberty*, p. 70, Univ. of Chicago., Chicago.
PRICE, D. J. (1959) "Contra Copernicus: a critical re-estimation of the mathematical planetary theory of Ptolemy, Copernicus and Kepler", in Clagett (1959).
PROWE, L. (1883) *Nicolaus Copernicus*, Berlin, 2 vols.; reprinted Osnabrück, Germany, 1967.
PURTILL, R. L. (1967) "Discussion: Kuhn on scientific revolutions" *Philosophy of Science: Official J. Philos. Sci. Soc.* **34**, no. 1, 53.
RAPPORT, S. and WRIGHT, H. (1965) *Astronomy*, Washington Square, N.Y.
ROSEN, E. (1959) *Three Copernican Treatises*, Dover, New York.
ROSEN, E. (1971) "Copernicus", in *Dictionary of Scientific Biography*, p. 401, Scribner's, New York.
RUSSELL, B. (1959) *A History of Western Philosophy*, Simon & Schuster, New York.
SANDAGE, A. (1961) *The Hubble Atlas of Galaxies*, Carnegie Institute, Washington, D.C.
SARTON, G. (1953) "The quest for truth: a brief account of scientific progress during the Renaissance", from *The Renaissance: A Symposium*, p. 40, Metropolitan Museum of Art, New York.
SCOTT, D. (1954) "The acceptance of the Copernican system in England", *Leaflets of Astron. Soc. Pacific*, no. 304.
SHANE, C. D. (1973) Letter to R. Berendzen, February 10.
SHAPERE, D. (1964) "The structure of scientific revolutions", *The Philosophical Review*, **73**, 383.
SHAPLEY, H. (1920) Letter to H. N. Russell, Shapley Collection, Harvard Archives, Cambridge.
SHAPLEY, H. (1958) *Of Stars and Men*, Beacon, Boston.
SHAPLEY, H. (1963) *The View From a Distant Star*, Basic Books, N.Y.
SHAPLEY, H. (1969) *Through Rugged Ways to the Stars*, Scribner's, N.Y.
SHAPLEY, H. and CURTIS, H. D. (1921) "The scale of the universe", *Bulletin National Research Council*, **2**, 171.
SIEMANKOWSKI, F. T. and MACKNIGHT, F. C. (1971) "Spatial cognition: success prognosticator in college science courses", *J. College Sci. Teaching*, **1**, no. 1, p. 56.
STEINBERGER, R. (1971) Letter, *Science*, **172**, 111.
STRUVE, O. (1942) "The work of Copernicus and the structure of the universe", *Bull. Polish Inst. Arts and Sciences in America*, **1**, 731.
STRUVE, O. and ZEBERGS, V. (1962) *Astronomy of the 20th Century*, Macmillan, New York.
Times, N.Y. (1924) "E. Hubble confirms that spirals are stellar systems," Nov. 23, p. 6.
Times, N.Y. (1930) Letter to the editor by B. Wright, Mar. 9, III. 5:6.
Times, N.Y. (1938) "The accident of life", Aug. 14, IV, 8:3.
TOULMIN, S. (1967) "Conceptual revolutions in science", *Synthese: Int. J. Epistemology, Methodology, and Philos. of Sci.* **17**, 75.
TOULMIN, S. and GOODFIELD, J. (1961) *The Fabric of the Heavens*, p. 179, Harper & Brothers, New York.
WRIGHT, H., WARNOW, J. N. and WEINER, C. (1972) *The Legacy of George Ellery Hale*, p. 46, MIT Press, Cambridge.

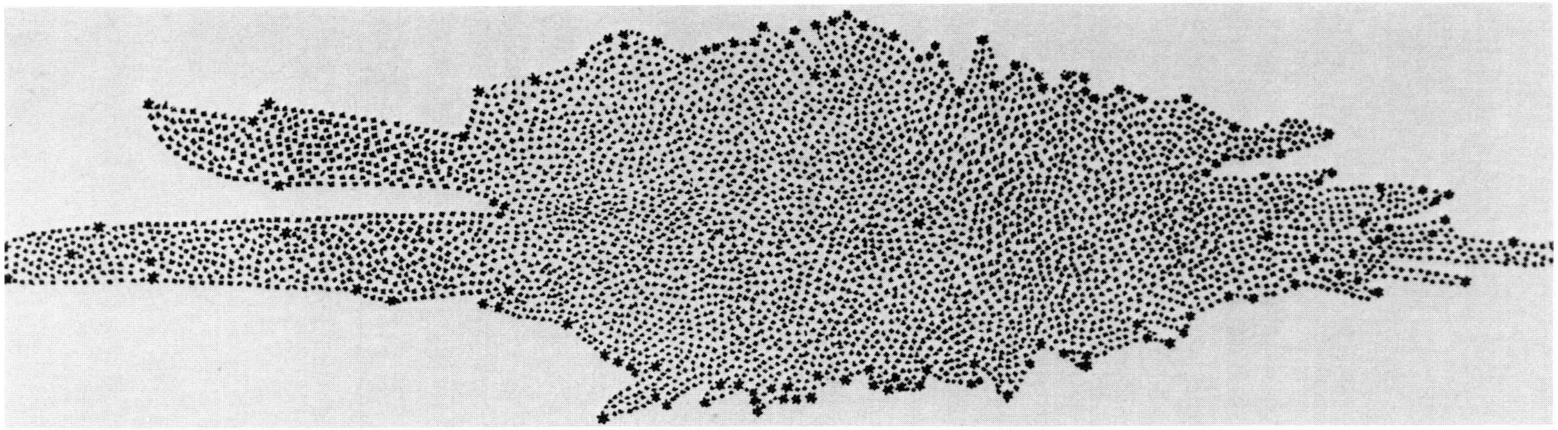

FIG. 46. William Herschel's model of the Milky Way, 1785, derived from "star gauges". The Sun's position is indicated by the large star near the center. (From M. A. Hoskin, *William Herschel and the Construction of the Heavens*, W. W. Norton Co., New York, 1963, p. 96.)

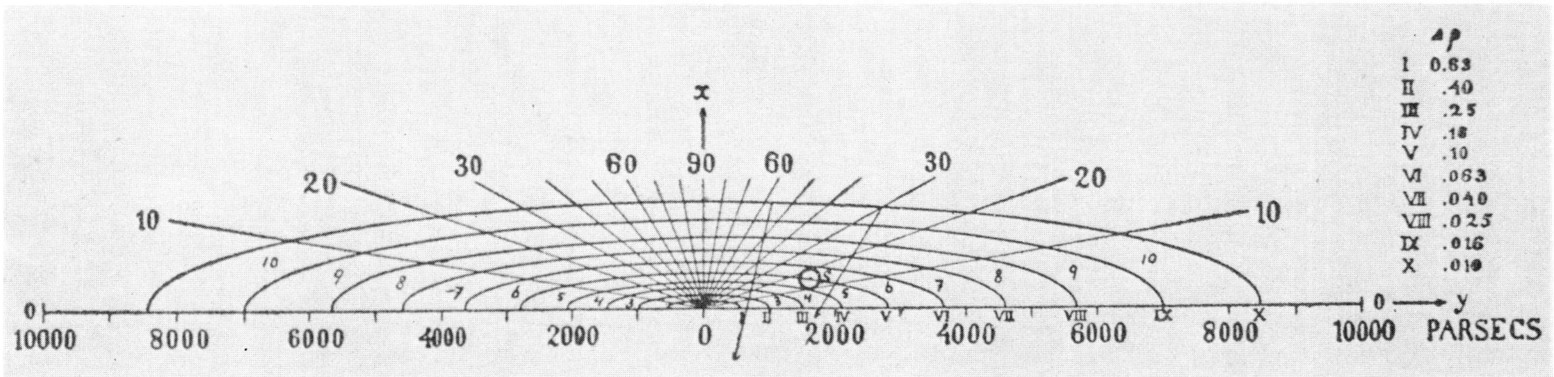

FIG. 47. The Kapteyn Universe, showing a cross-section of half the Milky Way. The Sun's position is indicated by an S. (From the *Astrophysical Journal* 55, 304, 1922.)

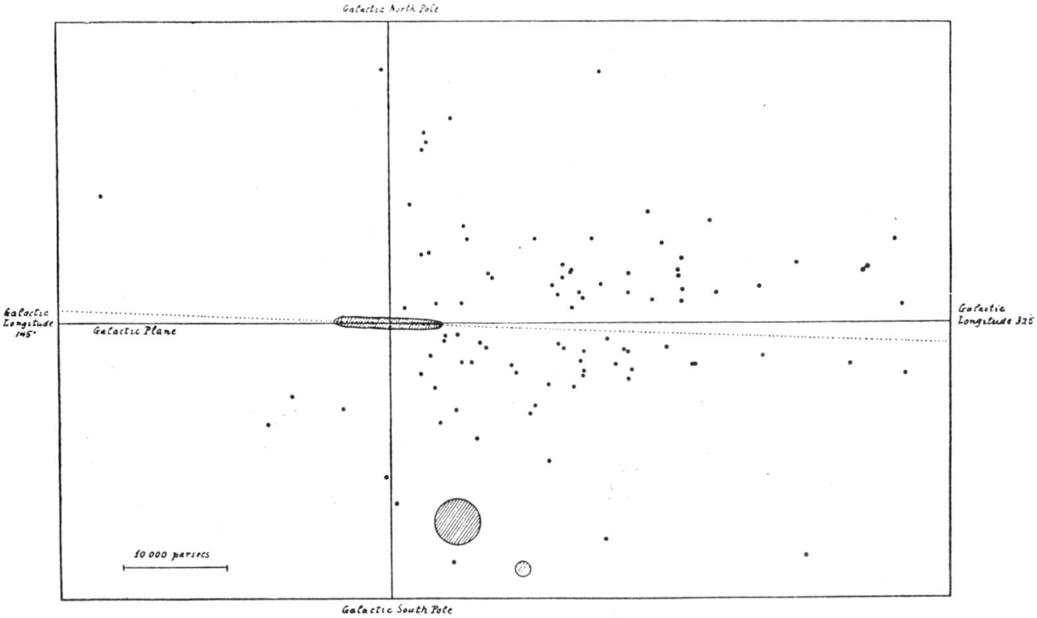

FIG. 48. The Shapley Universe, as depicted by R. J. Trumpler. The central object represents a cross-section of the Milky Way (the Kapteyn Universe). The full dots are globular clusters, and the shaded circles are the Magellanic Clouds. (From *Lick Observatory Bulletin XIV*, no. 420, p. 186, 1930.)

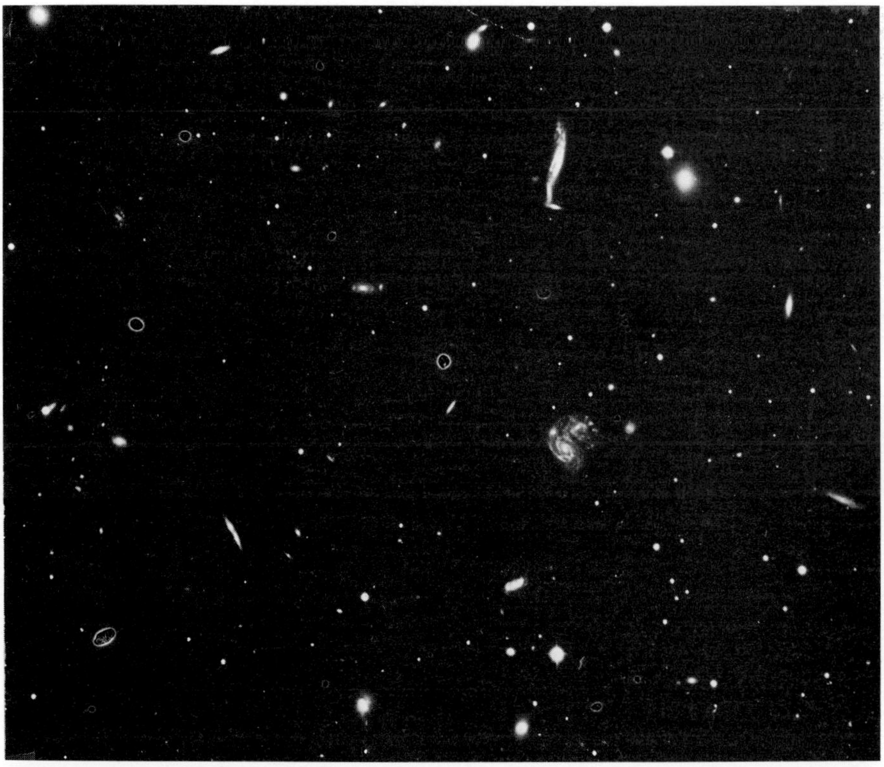

FIG. 49. The "Acentric Universe". A cluster of galaxies in the Constellation Hercules, photographed with the 200-inch telescope. (Courtesy of Hale Observatories.)

"Crisis" versus Aesthetic in the Copernican Revolution

Owen Gingerich

Smithsonian Astrophysical Observatory and Harvard College Observatory

In a chapter in *The Structure of Scientific Revolutions* entitled "Crisis and the Emergence of Scientific Theories", Thomas Kuhn states: "If awareness of anomaly plays a role in the emergence of phenomena, it should surprise no one that a similar but more profound awareness is prerequisite to all acceptable changes of theory. On this point historical evidence is, I think, unequivocal. The state of Ptolemaic astronomy was a scandal before Copernicus' announcement."[1] A paragraph later he elaborates:

> For some time astronomers had every reason to suppose that these attempts would be as successful as those that had led to Ptolemy's system. Given a particular discrepancy, astronomers were invariably able to eliminate it by making some particular adjustment in Ptolemy's system of compounded circles. But as time went on, a man looking at the net result of the normal research effort of many astronomers could observe that astronomy's complexity was increasing far more rapidly than its accuracy and that a discrepancy corrected in one place was likely to show up in another.

The existence of an astronomical crisis facing Copernicus in the early 1500s is presupposed by one author after another; perhaps it is most vividly expressed by de Vaucouleurs, who writes:

> The Ptolemaic system was finally overthrown as a result of the complexity which arose when an ever-increasing number of superimposed circles had to be postulated in order to

[1] Thomas S. Kuhn, *The Structure of Scientific Revolutions*, pp. 67–68, Chicago, 1962.

represent the ever-multiplying inequalities in the planetary motions revealed by observational progress.[2]

Nevertheless, my own researches have convinced me that this supposed crisis in astronomy is very elusive and hard to find, at least in the places where we are normally told to look. As a simple but powerful example of what I have in mind, let me cite the work of two leading ephemeris-makers of the sixteenth century, Johannes Stoeffler and Johannes Stadius.

Stoeffler was born in 1452. When late in life he became professor of mathematics at Tübingen, he already enjoyed a virtual monopoly with the ephemerides prepared by himself and Jacob Pflaum; these had continued through 1531 those of Regiomontanus. At Tübingen he extended his calculations to 1551, and these were published there posthumously in 1531.[3]

At about the same time that Stoeffler died (1530), Johannes Stadius was born (1527). In the 1560s Stadius taught mathematics at Louvain, and later he worked in Paris. Stadius was the first computer to adopt the Copernican parameters for a major ephemeris.[4] His own tables were, in effect, the successors to Stoeffler's, and their users included Tycho Brahe.

In this modern age of refined planetary theory and of electronic computers, it has become possible to calculate with fair precision where the planets really were in the sixteenth century, and hence I have been able to graph the errors in the planetary positions predicted by Stoeffler and by Stadius. These error patterns are as distinctive as fingerprints and reflect the characteristics of the underlying tables. That is, the error patterns for Stoeffler are different from those of Stadius, but the error patterns of Stadius closely resemble those of Maestlin, Magini, Origanus, and others who followed the Copernican parameters (see Figs. 50 and 51 on p. 94).[5]

The first result of this comparison is the fact that the errors reach approximately the same magnitude before and after Copernicus. In the Regiomontanus and Stoeffler ephemerides, the error in longitude for Mars is sometimes as large as 5°. However, in 1625, the Copernican errors for Mars reached nearly 5°, as Kepler complained in the Preface to his *Rudolphine Tables*.[6] And in Tycho's observation books, we can see occasional examples where the older scheme based on the *Alfonsine Tables* yielded better predictions than could be obtained from the Copernican *Prutenic Tables*. Now if the scandalous crisis of Ptolemaic astronomy was its failure to predict planetary positions accurately, Urania was left with nearly as much of a crisis on her hands after Copernicus.

Many simple historical accounts of the Copernican revolution emphasize not the accuracy but the simplicity of the new system, generally in contrast to the horrendous

[2] Gérard de Vaucouleurs, *Discovery of the Universe*, pp. 32–33, London, 1957.

[3] See Note A, pp. 91–92.

[4] See Note B, p. 92.

[5] See Note C, p. 92.

[6] See Note D, pp. 92–93.

complex scheme of epicycles-upon-epicycles supposedly perpetrated by pre-Copernican astronomers. This tale reaches its most bizarre heights in a recent *Encyclopædia Britannica*,[7] where the article on astronomy states that by the time of Alfonso in the thirteenth century, forty to sixty epicycles were required for each planet! More typically, we find what Robert Palter has called the "80–34 syndrome"—the claim that the simpler Copernican system required only thirty-four circles in contrast to the eighty supposedly needed by Ptolemy.[8] The Copernican count derives from the closing statement of his *Commentariolus*: "Altogether, therefore, thirty-four circles suffice to explain the entire structure of the universe and the entire ballet of the planets."[9] By the time Copernicus had refined his theory for his more mature *De revolutionibus*, he had rearranged the longitude mechanism, thereby using six fewer circles, but he had added an elaborate precession–trepidation device as well as a more complicated latitude scheme for the inner planets. Even Copernicus would have had difficulty in establishing an unambiguous final count.[10] A comparison between the Copernican and the classical Ptolemaic system is more precise if we limit the count of circles to the longitude mechanisms for the (Sun), Moon, and planets: Copernicus requires 18, Ptolemy 15.[11] Thus, the Copernican system is slightly more complicated than the original Ptolemaic system.

The 80–34 myth claims that the original simplicity of the Ptolemaic system was lost over the course of the ensuing centuries. "Theory patching was the order of the day", writes one recent author. The eighty circles presumably resulted from the piling of one epicycle on another, reminiscent of the lines

Great fleas have little fleas
  upon their backs to bite 'em,
And little fleas have lesser fleas,
  and so *ad infinitum*.

[7] See Note E, p. 93.

[8] Robert Palter, "An approach to the history of early astronomy", *History and Philosophy of Science*, **1**, 93–133 (1970). Palter traces the 80–34 myth back as far as Arthur Berry's *A Short History of Astronomy* (London, 1898).

[9] Edward Rosen, "Nicholas Copernicus, a biography", in his *Three Copernican Treatises*, 3rd ed., p. 90 (New York, 1971).

[10] According to Ernst Zinner, *Entstehung und Ausbreitung der Copernicanischen Lehre*, pp. 186–7 (Erlangen, 1943), Copernicus should have included precession, the regression of the lunar nodes, and the change of solar distance in his count in the *Commentariolus*, thus getting a total of thirty-eight circles. Arthur Koestler in *The Sleepwalkers*, pp. 572–3 (London, 1959), attempted to count the circles in *De revolutionibus*, but he overlooked the fact that Copernicus had by then replaced the so-called Ṭūsī couple in the longitude mechanisms by an eccentric, thereby listing at least six unnecessary circles; on the other hand, he could have claimed that the motion of the apsidal lines for Mercury and the superior planets each required a circle.

[11] Copernicus replaced the Ptolemaic mechanism for varying the site of Mercury's orbit with a Ṭūsī couple, and he also accounted for the apsidal motion of the Earth's orbit with two circles. If the apsidal motions for Mercury and the superior planets are counted, then Copernicus required twenty-two circles for the motions in longitude.

Astronomers have been fond of this view, because of the parallel between epicycles-on-epicycles and an analysis by Fourier series.[12] Nevertheless, this contrast between the simplicity of the Copernican system and the complexity of the detailed Ptolemaic mechanisms proves to be entirely fictitious.

Consider Stoeffler once more, the successor of Regiomontanus and the most successful ephemeris-maker of his day. If improvements were available in a patched-up scheme of epicycles-on-epicycles, surely Stoeffler would have used them. Two extensive sets of calculations allowed me to investigate this possibility.

First, I recomputed the thirteenth-century *Alfonsine Tables*, showing that they are based on a pure Ptolemaic theory, that is, with an eccentric, equant and single epicycle for the superior planets. The parameters were almost all identical to those originally adopted by Ptolemy, but the precessional motion had been augmented by trepidation, an improvement irrelevant to the discussion of epicycles-on-epicycles. Second, I used the *Alfonsine Tables* to generate a daily ephemeris for three centuries;[13] these positions agreed so closely with those published by Stoeffler that I am forced to conclude he used the unembellished Ptolemaic system, as transmitted through the *Alfonsine Tables* (Fig. 52, p. 95).[14]

Thus, this second result of investigating the ephemerides indicates that only a simple, classical Ptolemaic scheme was used for the prediction of planetary positions in 1500. I am convinced that the complex, highly embroidered Ptolemaic system with all the added circles is a latter-day myth. To support my view, there are at least two more good arguments, although I can mention them only in passing. First, the most sophisticated understanding of the Ptolemaic system in the fifteenth century is reflected in the tract against Cremonensis,[15] in which Regiomontanus picks faults with an anonymous Medieval work, the *Theorica Planetarum*. One receives the impression here that in 1464 astronomers were once again just able to comprehend Ptolemy, but scarcely able to improve on his work. Second, the astonishing, almost

[12] A recent letter to *Physics Today* (24, p. 11) remarked that 400 years ago the *Physical Review* might have been full of such papers as "A ten epicycle fit to the orbit of Mars", and a review article on radio galaxies in *The Astronomical Journal*, 77, p. 541, 1972, summarized with "The question is, 'Are we drawing too many epicycles?' "

[13] E. Poulle and O. Gingerich, "Les positions des planètes au moyen âge: Application du calcul électronique aux tables Alphonsines", *Comptes Rendus de l'Académie des Inscriptions et Belles Lettres*, pp. 531–48, 1968.

[14] Recently, I found in the Badische Landesbibliothek in Karlsruhe what I believe to be Stoeffler's personal manuscript copy of these tables, which he may have used in calculating his ephemerides. It is Codex Ettenheim-Münster 33, 93r–198r. I wish to thank the director, Dr. Kurt Hannemann, for showing me this manuscript. See Karl Preisendanz, *Die Handschriften des Klosters Ettenheim-Münster*, IX in *Die Handschriften der Badischen Landesbibliothek in Karlsruhe*, Wiesbaden, 1932.

[15] Johannes Regiomontanus, *Disputationes contra Cremonensia deliramenta*, Nuremberg, 1474 or 1475. According to Ernst Zinner, *Leben und Wirken des Joh. Müller von Königsberg*, 2nd ed., p. 335 (Osnabrück, 1968), Regiomontanus wrote the tract in August 1464.

complete absence of recorded observations before 1450 again suggests that pre-Copernican astronomers had little basis for adding those mythical epicycles-on-epicycles. I simply cannot believe Kuhn's statement that "as time went on, a man looking at the net result of the normal research effort of many astronomers could observe that astronomy's complexity was increasing far more rapidly than its accuracy and that a discrepancy corrected in one place was likely to show up in another".[1]

I am willing to grant that Copernicus' cosmology represents, in a certain profound sense, a simplification, but I refuse to concede that the Ptolemaic theory had by the beginning of the sixteenth century reached a complex, patched-up state nearing collapse. In terms of the detailed mechanism for any particular planet, it would have been very difficult for Copernicus' contemporaries to distinguish between the two schemes on the basis of complexity.

Where, then, is the astronomical crisis that Copernicus faced? Kuhn goes on to say:

> By the early sixteenth century an increasing number of Europe's best astronomers were recognizing that the astronomical paradigm was failing in application to its own traditional problems. That recognition was prerequisite to Copernicus' rejection of the Ptolemaic paradigm and his search for a new one. His famous preface still provides one of the classic descriptions of a crisis state.[16]

This preface is the last extant piece of Copernican prose, written just before the publication of his book. A polemical passage, it attempts to justify his radical departure from traditional cosmology and to protect his work from future detractors. If one believes astronomy was at the point of crisis, then it is perhaps possible to read it as a classic description of a crisis state.

On the other hand, I believe that an alternative reading is preferable. After criticizing the alternative system of homocentric spheres, and indirectly, Ptolemy's equant, Copernicus says:

> Nor have they been able thereby to discern or deduce the principal thing—namely the design of the universe and the fixed symmetry of its parts. With them it is as though one were to gather various hands, feet, head and other members, each part excellently drawn, but not related to a single body, and since they in no way match each other, the result would be monster rather than man.[17]

This "fixed symmetry of its parts" refers to the fact that, unlike in the Ptolemaic scheme, the relative sizes of the planetary orbits in the Copernican system are fixed with respect to each other and can no longer be independently scaled in size. This is certainly one of the most striking unifications brought about by the Copernican

[16] Kuhn, *op. cit.*, p. 69.

[17] N. Copernicus, *De revolutionibus orbium coelestium*, f. iii(v), Nuremberg 1543. Edward Rosen suggests for "certam symmetriam" the term "true symmetry". I believe that "fixed" conveys a slightly better nuance in this context.

system—what I would call a profound simplification. Clearly, this interlinking makes the unified man, and in contrast the individual pieces of Ptolemy's arrangement become a monster.

What has struck Copernicus is a new cosmological vision, a grand aesthetic view of the structure of the Universe. If this is a response to a crisis, the crisis had existed since A.D. 150. Kuhn has written that the astronomical tradition Copernicus inherited "had finally created a monster", but the cosmological monster had been created by Ptolemy himself.

In this view, there is no particular astronomical reason why the heliocentric cosmology could not have been defended centuries earlier, and it is in fact shocking that Copernicus, with the accumulated experience of fourteen more centuries, did not come up with a substantial advance in predictive technique over the well-honed mechanisms of Ptolemy. The debased positivism that has so thoroughly penetrated our philosophical framework urges us to look to data as the foundation of a scientific theory, but Copernicus' radical cosmology came forth not from new observations but from insight. It was, like Einstein's revolution four centuries later, motivated by the passionate search for symmetries and an aesthetic structure of the universe. Only afterward the facts, and even the crisis, are marshalled in support of the new world view.[18]

But why, if all this is true, did a Copernicus come in the sixteenth century, and not in the fourteenth or even the tenth century? Were the astronomical questions in Krakow in 1492 particularly conducive to challenging the old order? I have no doubt but that the growing problems of precession, trepidation, and the motion of the eighth sphere acted as a spur to Copernicus' thinking about astronomy. His attack on this problem demonstrates his unusual level of technical ability, which had certainly been rare in the Middle Ages. Copernicus' examination of precession may have led him to consider a moving Earth.[19] Nevertheless, the heliocentric system is scarcely a necessary consequence of the observation of precession.

No, I believe that it was something outside astronomy in the European intellectual climate in the sixteenth century that set the stage for the introduction of a new paradigm; as Professor Benjamin Nelson put it in an earlier paper in this symposium —it had something to do with "societies, communities, and communications". In his words, the flowering of new world views must be considered within the context of complex sociocultural structures. The sixteenth century was manifestly an age of

[18] See Gerald Holton, "Einstein, Michelson, and the 'Crucial Experiment' ", *Isis*, **62**, 133–97 (1969).

[19] J. R. Ravetz, in *Astronomy and Cosmology in the Achievement of Nicolaus Copernicus* (Wrocław, 1965), argues that studies of precession may have led to the Copernican cosmology. L. Birkenmajer, in *Mikolaj Kopernik* (Krakow, 1900), suggested that the deficiencies in the Ptolemaic lunar model may have started Copernicus on the road to the heliocentric system. Important as these may have been in the development of Copernicus' technical proficiency, there is no convincing argument that these studies would have led to a Sun-centered cosmology.

change. While Copernicus was a student at Krakow, Columbus set sail across an unknown ocean. The new explorations made Ptolemy's time-honored geography obsolete. Discoveries of classical authors brought in a new humanism with fresh Neoplatonic ideals. Even the traditional authority of the Church was to crumble before the challenge of Luther and the reformers.

A powerful catalyst for these changes was the explosive proliferation of printing.[20] As a student in Krakow, Copernicus could secure and annotate his own printed set of *Alfonsine Tables* as well as Regiomontanus' *Ephemerides*. Later, probably in Italy, he obtained Regiomontanus' *Epitome of Ptolemy's Almagest*; the close paraphrases of many of its passages in the *De revolutionibus* show the formative role this book played in his researches. Still later, the first full printed *Almagest* of 1515 provided another useful source of data.[21] Ultimately, it was the *printed* edition of his *De revolutionibus* that prevented his ideas from falling into oblivion.

In many ways, the world was ready for an innovative view of the cosmos. Copernicus, with both the intellect and the leisure to fashion a new cosmology, arrived on the scene at the very moment when the increased flow of information could both bring him the raw materials for his theory and rapidly disseminate his own ideas. An imaginative thinker striving to uncover fresh harmonies in the universe, he also achieved the technical proficiency to command respect for his mathematics and his planetary tables. One can easily argue that Copernicus was not the equal of Ptolemy or of Kepler in mathematics, although for his day he stood well above his contemporaries. Yet as a sensitive visionary who precipitated a scientific revolution, Copernicus stands as a cosmological genius with few equals. In celebrating his birth, we celebrate the man who, perhaps unwittingly, is the founder of modern science.

Notes

NOTE A. In 1474 in Nuremberg, Regiomontanus printed his own ephemerides for 1475 through 1506, and these were reissued by various printers, including Ratdolt in Venice. Stoeffler and Pflaum issued their ephemerides in Ulm in 1499 for the years 1499 to 1531, with the title *Almanach nova plurimis annis venturis inservientia*, and these were repeatedly reissued by Liechtenstein in Venice. I have not yet ascertained if they recalculated the overlapping period from 1499 to 1506. Stoeffler's 1531 edition in Tübingen, with the title *Ephemeridum opus*, was edited by the successor to his professorial chair, Phillip Imsser; these tables were also promptly reprinted by Liechtenstein in Venice.

---

[20] See Owen Gingerich, "Copernicus and the impact of printing" on pp. 201–20 of this volume. See also E. L. Eisenstein, "The advent of printing and the problem of the Renasissance", *Past and Present*, no. 45, pp. 19–89 (1972).

[21] A detailed discussion of Copernicus' use of these books is found in L. Birkenmajer, *Mikolaj Kopernik* (Krakow, 1900). A useful list of books owned by, or available to, Copernicus is found in L. Jarzębowski's *Biblioteka Mikolaja Kopernika* (Toruń, 1971). An earlier list, of the Copernican books now found in Sweden, is E. Barwiński, L. Birkenmajer, and J. Łos, *Sprawozdanie z Poszukiwań w Szwecyi*, pp. 94–119 (Krakow, 1914).

Edward Sherburne gives a charming account of the death of Stoeffler in the biographical appendix to his *The Sphere of Marcus Manilius* (London, 1675), p. 46:

> His death, or the occasion thereof at least, was very remarkable (if the Story be True). Having found by calculation, that upon a certain Day his life was like to be endangered by some ruinous accident, and the day being come, to divert his thoughts from the apprehension of the danger threatening him, he invites some Friends of his into his Study, where, after discourse, enticing into some dispute, he, to decide the controversie reaches for a Book, but the Shelf on which it stood being loose came down with all the Books upon him, and with its fall so bruised him, that he died soon after of the hurt, *Voss. in Addend. ad Scient. Mathemat.* But the whole Story of his Death, of which some make Calvisius the Author, is false by the Testimony of *Jo. Rudolphus Camerarius Genitur. 69. Centur. 2.* who had it from *Andraas Ruttellius* his Auditour; for he died of the Plague at *Blabira Feb.* 16. 1531 in the 78th year of his Age, happening (according to Calculation if you will believe it) from the Direction of ☉ to ♂.

NOTE B. Copernicus' own almanac was never printed and is now lost (see Edward Rosen, "Nicholas Copernicus, a biography", in his *Three Copernican Treatises*, pp. 374–5, 3rd ed., New York, 1971). Rheticus published an ephemeris for a single year, 1551, based on the tables in *De revolutionibus.* E. Reinhold published an ephemeris for 1550 and 1551, using his Copernican-based *Tabulae Prutenicae* (Tübingen, 1551); subsequent workers generally adopted Reinhold's tables as their avenue to the Copernican parameters. Stadius' *Ephemerides novae* (Cologne, 1556) included predictions for 1554–70, and later editions carried the tables through 1600. A posthumous edition went to 1606, but the additional years were probably appended by the publisher from the Alfonsine-based ephemerides of Leovitius. Stadius published his own planetary tables, *Tabulae Bergenses aequabilis et apparentis motus orbium coelestium* (Cologne, 1560), but these were essentially a plagiarism of the *Tabulae Prutenicae.* Lynn Thorndike (*A History of Magic and Experimental Science*, vol. 5, pp. 303–4, New York, 1941) quotes Tycho Brahe's estimate of Stadius as having been "more facile than accurate", an opinion apparently shared by Maestlin and Magini, who eventually produced major alternative ephemerides of their own.

NOTE C. The ephemerides used for the figures are Johannes Stoeffler *op. cit.*; Cyprian Leowitz, *Ephemeridum novum atque insigne opus ab anno 1556 usque in 1606 accuratissime supputatum*, Augsburg, 1557; Johannes Stadius, *Ephemerides novae et epactae ab anno 1554 ad annum 1600*, Cologne, 1570; Michael Maestlin, *Ephemerides novae ex tabulis Prutenices anno 1577 ad annum 1590 supputatae*, Tübingen, 1580; G. A. Magini, *Ephemerides coelestium motuum secundum Copernici observationes supputatae*, Venice, 1582. The comparisons were made against the computed longitudes in Bryant Tuckerman, *Planetary, Lunar, and Solar Positions A.D. 2 to A.D. 1649, Memoirs of the American Philosophical Society*, vol. 59, Philadelphia, 1964. Both figures were prepared by Barbara L. Welther.

Additional error graphs from sixteenth- and seventeenth-century ephemerides can be found in Owen Gingerich, "The theory of Mercury from Antiquity to Kepler", *Actes du XII Congrès International d'Histoire des Sciences*, vol. IIIA, pp. 57–64, 1971, and "Kepler's place in astronomy", *Vistas in Astronomy*, vol. 18, ed. A. and P. Beer, pp. 261–78, 1974.

NOTE D. "Johannes Kepler: Preface to the Rudolphine Tables", translated by Owen Gingerich and William Walderman, *Quarterly Journal of the Royal Astronomical Society*, 13, pp. 360–73, 1972; see especially p. 367.

Tycho frequently compared his own observations to the predictions from the Alfonsine and Copernican tables, usually to the advantage of Copernicus. A particularly favourable comparison occurred at the time of the great conjunction of Jupiter and Saturn in 1583 (see Fig. 51), although by 20 August, 1584, Tycho's comparison for Jupiter showed the two schemes equally in error, and by 21 December, 1586, the Alfonsine calculation was decidedly better, especially in latitude. Frequently, the Copernican latitudes proved inferior, even when the longitude excelled—for example, for Saturn on January 24, 1595. Tycho compared lunar positions in December 1594, and toward the end of the month the Alfonsine-based Leovitius ephemeris was superior. The most conspicuous

Copernican errors found by Tycho occurred during the August opposition of Mars in 1593, exceeding 5°; this configuration repeated in 1625 when Kepler noted the large errors during the particularly close approach of Mars. Tycho's investigations are published in J. L. E. Dreyer (ed.), *Tychonis Brahe Dani Opera Omnia*, **10–13**, Copenhagen, 1923–6.

NOTE E. "Astronomy. I. History of astronomy. B. Mediaeval astronomy", *Encyclopædia Britannica*, vol. 2, p. 645, Chicago, 1969:

> King Alfonso X of Castile kept a number of scholars occupied for ten years constructing tables (the Alphonsine tables, *c*. 1270) for predicting positions of the planetary bodies. By this time each planet had been provided with from 40 to 60 epicycles to represent after a fashion its complex movement among the stars. Amazed at the difficulty of the project, Alfonso is credited with the remark that had he been present at the Creation he might have given excellent advice. After surviving for more than a millennium, the Ptolemaic system had failed; its geometrical clockwork had become unbelievably cumbersome and without satisfactory improvements in its effectiveness.

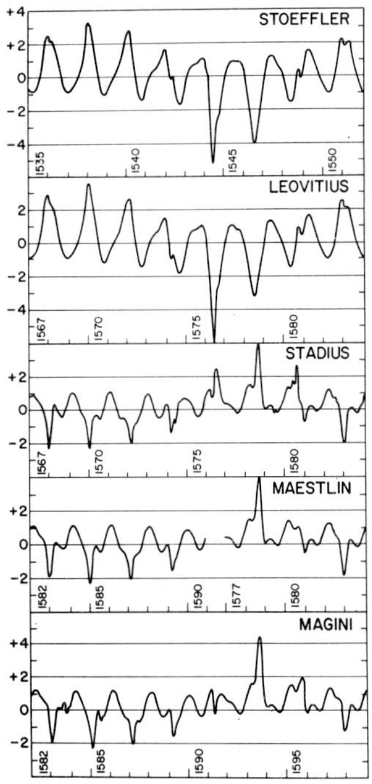

FIG. 50. The errors in the predicted longitude for Mars in the Alfonsine-based ephemerides of Stoeffler and Leovitius and three Copernican-based ephemerides. Some of, but not all, the typographical or obvious computation errors of Stadius have been corrected. Note the close agreement in error patterns after intervals of 15 and 32 years. (Drawn by Barbara L. Welther.)

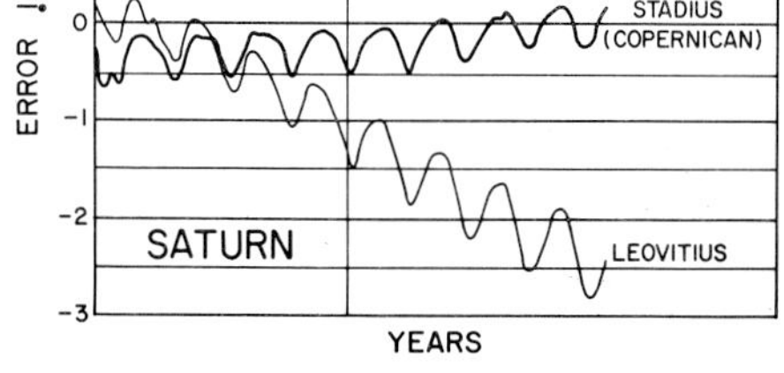

FIG. 51. Errors in predicted longitudes of Jupiter and Saturn near the time of their great conjunction in May 1583, a configuration closely observed by Tycho Brahe. There is evidently much computational noise in Stadius' positions for Jupiter. (Drawn by Barbara L. Welther.)

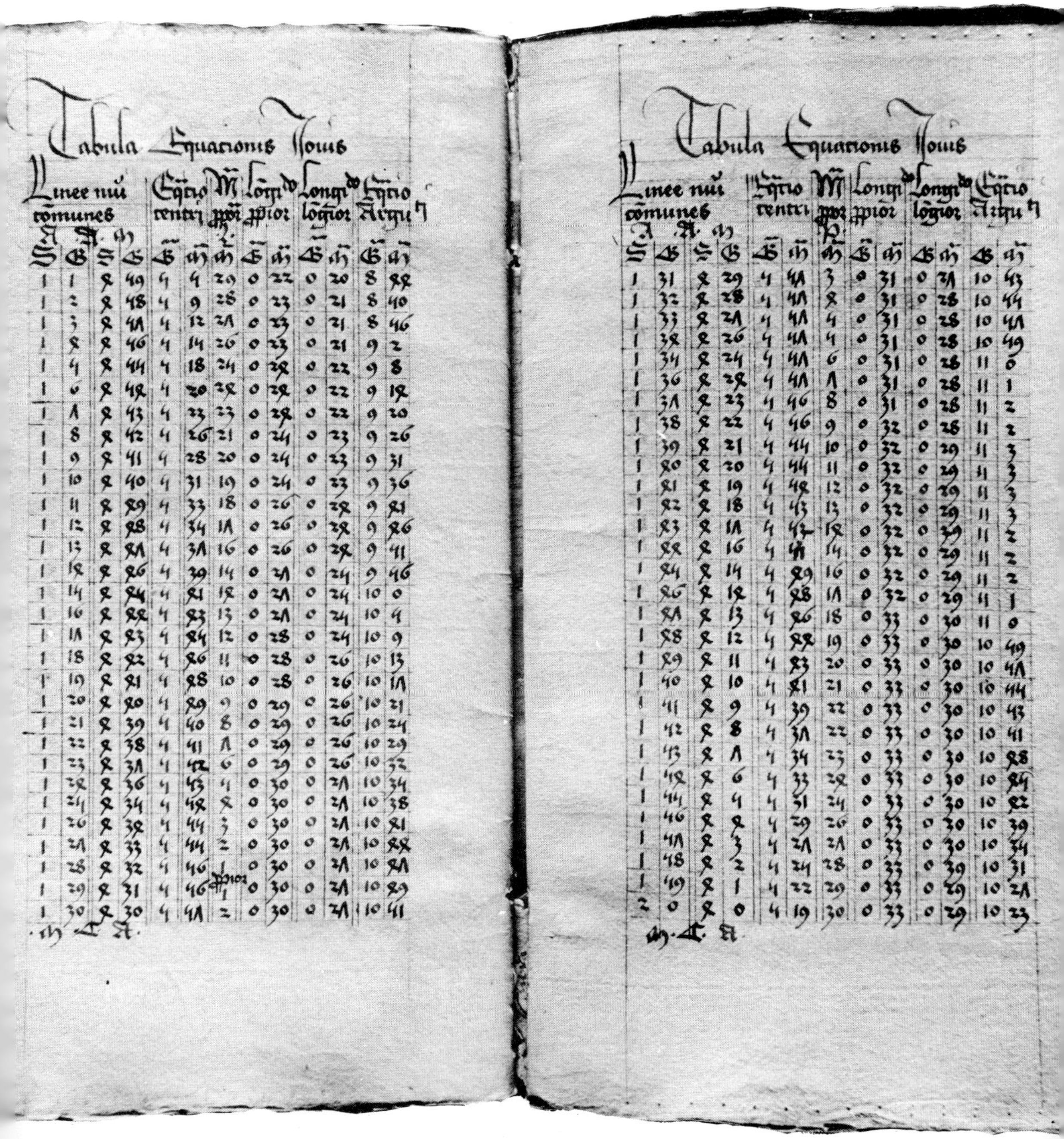

FIG. 52. A page from the *Alfonsine Tables*. (Photograph by Charles Eames.) Courtesy of the University Observatory Library, Uppsala.

Copernicanism as a Scientific Revolution

DUDLEY SHAPERE

University of Illinois, Urbana, Illinois

THERE are, in the history of science, episodes which seem to involve changes of such fundamental character and broad implications that we feel impelled to refer to them as "scientific revolutions". One of the primary aims of the historian and philosopher of science, especially in recent years, has been to try to gain an understanding of such major episodes of scientific change. (This is not, of course, to say that we are not also concerned with the analysis of more "normal" scientific change.)

The form this attempt has often taken has been to develop a general concept of "scientific revolution" which, while illuminating the logic of scientific reasoning, would do justice to at least what are considered to be paradigm cases of "scientific revolutions". These paradigm cases—the ones on which discussions of the concept have tended to focus—include "the Copernican Revolution", "the Galilean Revolution" (though sometimes these two are grouped together, along with other developments, under the title, "the Scientific Revolution"), "the Chemical Revolution" (of Lavoisier and/or Dalton), "the Darwinian Revolution", "the Einsteinian Revolution" (referring almost invariably to the theory of special relativity), and "the Quantum Revolution" (usually either Bohr's of 1913, or that of Schrödinger, Heisenberg, and others of the middle and late 1920s—or both).

But this approach suffers from several defects. For one thing, the list of supposedly paradigm cases which usually serve as the bases for generalization embodies a number of potentially severe biases. The "scientific revolutions" commonly examined still tend to be chiefly, if not almost exclusively, ones from the physical sciences: the above list includes only one case from biology. Perhaps, in trying to generalize about scientific revolutions, we should explicitly structure our analyses to account also for a "Mendelian Revolution" (referring not to Mendel's original work, but to the developments stemming from its recovery by de Vries, Correns and Tschermak in 1900), a "Huttonian Revolution", and a "Freudian Revolution", to mention only three. Again, the usual list of "scientific revolutions"—the episodes on the basis of which analyses of that concept are framed, and to which they are expected to do justice—tends to include only episodes which, when viewed from a modern perspective, led to success—that is, views which eventuated in our own. But there were other episodes which, though ultimately "failures" or "blind alleys" from our viewpoint, nevertheless brought about profound changes in the character of science. Historians, to be sure, have investigated, or begun to investigate (though not always with the greatest possible philosophical sophistication), what we might refer to as the "Aristotelian Revolution", the "Buridanian Revolution", the "Paracelsian Revolution", and the "Stahlian Revolution". But philosophers—and also, historians, when they are thinking on a loftier conceptual level about the "structure of scientific revolutions"—tend to focus their analyses of the abstract concept around the list I originally gave above.

Furthermore, the list of paradigm cases, even as amended, ends with the late 1920s; but from the scientific viewpoint, at least, the frequency of "revolutionary" developments has accelerated more than geometrically in the last half century. Each decade has witnessed spectacular alterations in the fundamental structure of at least one area of science: nuclear physics had its heyday in the 1930s; the post-war years saw rapid and profound innovation in a number of areas of physics. The 1950s ushered in an age of molecular biology which revolutionized the subject of genetics. In the 1960s immunology built rapidly toward a theoretical unity which it had previously lacked; geology has been doing the same beginning in the late 1960s. And astronomy, undergoing perhaps the most exciting and explosive development of its long history in the 1960s and early 1970s, now threatens to reach a new profound alteration of basic ideas about the Universe—perhaps dragging physics along with it in the throes of revolution—when the distance and energy problems of quasars and radio galaxies are resolved (along with presumably associated problems about black holes, galactic nuclei, etc.). Perhaps it is too early yet to speak of, or at least to assess, a Yukawa, or a Watson–Crick, or a Burnet–Medawar, or a Vine–Hess, or a B^2FH Revolution; but if the concept of a scientific revolution is to have any explanatory role in the attempt to understand the scientific enterprise, surely it must be applicable to such cases as well as to those traditionally so christened. The very rapidity of occurrence of such "revolutions" now is something that can be ignored only at the peril of a philosophy of science.

Besides these limitations in the list of paradigmatic examples usually considered, approaches to the problem of the analysis of major scientific change must also be wary of the examples themselves: there is a constant need for the claims of those cases to being "revolutionary" to be reconsidered, partly in the light of new knowledge about the relevant historical situation, and partly in the light of sharpened understanding of the philosophical problems surrounding the notion of "revolution" in particular, and of scientific change more generally. Thus, much energy in recent years has been expanded in attempting to downgrade Galileo's work in the face of historical evidence; a more balanced view is by now emerging.

But the multitude and diversity of the claimants to the title of "scientific revolutions", coupled with their increasing frequency, raises further questions about the standard approach to the problem. As I suggested earlier, that approach has been to try to formulate a general concept of "scientific revolution" which will throw light on the logic of scientific reasoning, while being adequate to paradigm cases of major scientific change. The assumption, in other words, seems to be that *there is only one "logic" (general pattern of reasoning) common to all cases of major scientific change.* Such an assumption is embodied in the very idea that there is *a* (single) concept of "scientific revolution" to be analyzed; and it is reiterated in the way the problem is usually conceived: What is *the* structure of scientific revolutions?—as if there were only one. But might it not be that episodes of scientific change which are undeniably major are in many cases of very different character—so different, perhaps, that a common pattern can be found only by abstracting from the concreteness of the cases almost to the point of emptiness? Perhaps there are very different *kinds* of things—very different sorts of reasoning—involved in what are considered to be major scientific advances. (Perhaps, for instance, their character changes with the increasing sophistication of scientific reasoning.)

A second major assumption of many current treatments of major scientific change is that *this one "logic" assumed to be common to all scientific revolutions is different in fundamental ways from the sorts of reasoning employed in more "normal" cases of scientific development.* This assumption, too, seems to me highly questionable. In short, the very philosophical and historical utility of the concept of "scientific revolution" needs to be examined in the light of case analyses.

In attributing these two theses to contemporary approaches to the problem of scientific revolutions, I am being fairer to some of them than I think they really deserve. For some of those current views of the nature of scientific revolutions in effect *deny* that there is *any* "logic" or reason involved in a scientific revolution. Scientific revolutions, according to those views, constitute *complete* breaks with the past, redefining the criteria for scientific facts, problems, and solutions, and, in particular, redefining what constitutes a reason for accepting or rejecting a scientific proposition. According to such views, then, there *could* not be a reason for accepting a revolutionary view, since what counts as a reason is only determined by the making of that decision. Such a view in effect rejects the attempt to understand scientific change rather than illuminating it. I cannot accept that kind of view, for reasons which I will not enter into here;[1] let me say only that I think not only that there are as a matter of fact no cases satisfying this view of scientific revolutions as completely radical breaks with the past, but also that there *could* be no such cases, the view being the result of conceptual confusions rather than of historical analysis.

It will not be possible in this short paper to examine all the issues I have raised regarding the analysis of major scientific change—issues about the domain of cases to be considered in attempting to develop a philosophical theory of scientific revolutions, and about the utility of attempting to develop a single such account applicable to all such cases. What I will do is to examine some of these questions with regard specifically to the "Copernican Revolution".

It is often claimed that Copernicus' work constituted or produced a "scientific revolution" because, in removing man from the center of the Universe, it downgraded him from a position of major importance morally as well as physically. (This view is, of course, perfectly consistent with the fact that this is not at all what Copernicus or Galileo *aimed* to do.) This is undoubtedly true in many respects: a whole fabric in which the moral and the physical were intertwined collapsed through this "humbling" of man. (And that is why we do not, in *this* respect, speak also of a Shapleyan, or a Hubblean, or a Baadean Revolution, despite the fact that each of those steps decreased man's significance, as far as mere size and centrality of location are concerned, immeasurably beyond what Copernicus' move did: for, like breaking the 4-minute mile, the psychological barrier was obliterated by the first such step.)

But to the extent that this view of the "revolutionary" character of Copernicanism is true, it has nothing to do with the *scientific* aspects of the case: in so far as scientific considerations might have played a role in the historical events here, the fact that humanity was thus humbled is not the kind of consideration that could count in favor of or against the *scientific* promise or acceptability of the theory. I suggest therefore that this sense of the "Copernican Revolution" should be referred to as a *cultural* rather than as a *scientific* revolution. True, the cultural revolution ultimately produced an environment in which science could flourish; but the existence of such an environment is irrelevant to whether a certain view is scientifically promising or acceptable.

Thus, the question in which I am interested here is, in what sense, if any, did Copernicanism constitute a major *scientific* development, i.e. a *scientific* "revolution"? Did it have a major scientific impact because of the *problems* it dealt with (in so far as those problems, those motivations, can naturally be called scientific)? Or because of some radical characteristics of the theory itself—its new "concepts", for example?

[1] For some of the reasons, see my "The structure of scientific revolutions", *Philosophical Review*, pp. 383–94 (July 1964); "Meaning and scientific change", in R. Colodny (ed.), *Mind and Cosmos: Explorations in the Philosophy of Science*, pp. 41–85, Pittsburgh, University of Pittsburgh Press, 1966; "The paradigm concept", *Science* **172**, pp. 706–9 (May 14, 1971).

In attempting to assess the validity of the claim that Copernicus' work constituted a scientific revolution, we are immediately confronted with certain difficulties. In the first place, there is the well-known conservatism of Copernicus himself. His motive of restoring astronomy to the ancient purity of uniform circular motion, by eradicating the equant—to say nothing of his more mystical Neo-Platonic associations—was not exactly innovative. Surely the drive toward resolution of the dichotomy Einstein saw between mechanics and electrodynamics, between discrete and continuous views of nature, was more profoundly revolutionary than these motivations of Copernicus.[2] And as for the theory itself, it embodied the technical apparatus of traditional astronomy: it was, as he put it, "a more reasonable arrangement of circles", including epicycles—techniques familiar since the days of the Greek geometers. Surely Faraday's fields, or the quanta of Planck and Einstein, or Yukawa's exchange–particle interpretation of nuclear forces, constituted far more pervasive and fundamental theoretical alterations in science than did the mere interchange of position of two entities in a theory of the solar system, especially since that theory preserved the traditional theoretical, i.e. geometrical, devices.

Any truly "revolutionary" character of Copernicanism as a *scientific* revolution—any sense in which it was a major scientific development—thus seems to elude us when we look at the problems its author meant it to confront, or the conceptual apparatus of the theory itself. Could its revolutionary character (its scientific revolutionary character) reside, then, in its scientific *consequences*? It is sometimes claimed, for example, that the Copernican theory was important because it "called for" or "required" the rejection of Aristotelian physics and its replacement by a new physics. Indeed, the implication is sometimes stronger: that the Copernican theory called specifically for *Newtonian* physics. This view, however, is unacceptable, at least as it stands, for a number of reasons. In the first place, there was a long tradition in astronomy according to which astronomical theories were considered to be mere predictive devices, constructed to "save the phenomena", so that there was no need for them to be compatible with, or be supported by, *any* physics. And if this was to be the interpretation of the Copernican theory, there was no conflict with Aristotelian physics, no call for a new, or even for any, physics. Such conflict would arise only if the theory were taken as a realistic representation of the structure of the solar system (as Copernicus himself seems to have taken it), rather than as an instrument of calculation or prediction. But further, even when it is taken in a realistic sense, still, although it is true that Aristotelian physics—or, more accurately, certain components

[2] There are, however, even in the Einstein case, features of similarity to earlier views. Einstein seems to have regarded his early work on the quantum theory ("On a heuristic viewpoint concerning the production and transformation of light", *Annalen der Physik*, 17, 1905) as a reaction away from the continuum (wave) theory of light (or, more generally, of energy production) and a return to a particulate view having much in common with that of Newton. At the time, the latter was universally regarded as having been overthrown by the work of Young, Fresnel and others, culminating with the electromagnetic theory of Maxwell.

thereof—would have to be rejected, it is not strictly correct to say that a *new* physics would have to be constructed to replace it. On the contrary, there was a physics, or the makings of one, available already for a Copernican system: namely, the impetus theory, supplemented by the Platonic–Oresmean theory of gravity, according to which gravity (I include levity), rather than being a relationship between a physical entity and a spatial location (a "natural place"), was a relationship between like entities. The impetus theorists had already argued for the possibility of terrestrial axial rotation; and the "natural place physics" was not needed if earthy matter fell, not toward the center of the Universe, but toward other earth, as lunar or solar matter fell toward their respective likes. And from these arguments, it was an easy step to the possibilities of other sorts of terrestrial motion. (Indeed, arguments similar to all these are employed by Copernicus.) And even apart from the impetus theory, the "new" physical theories—all very different—with which Galileo, Kepler, and Descartes buttressed their Copernicanism were certainly not that of Newton. The only point that remains from this contention is that with, and only with, the adoption of a realistic interpretation of Copernican astronomy, *some* physics would be required which was non-Aristotelian in at least certain respects.[3]

None of this is to deny that Copernicanism deserves to be considered a case of major scientific innovation—that it deserves to be called "revolutionary". Though scholars now recognize that Copernicus' theory held no distinct advantage over Ptolemy's (or sophisticated versions thereof) with respect to observational adequacy, predictive accuracy, or number of epicycles, it did, in its details, provide a far more unified account of solar system astronomy than did its predecessors. The basic dichotomy between the accounts of inferior and superior planets, the abritrary dependence of deferential or epicyclic motion, and of the orientation of epicycles, on the (mean) Sun, the lack of a definite criterion for positioning Mercury and Venus in order out from the Earth, or for determining the relative sizes of the planetary orbits in general—all these, and several others, were characteristic disunities in the Ptolemaic theories which were removed by the Copernican theory; what had been arbitrary assumptions or mere coincidences were so no longer, but followed naturally from the theory.[4] (It is important to emphasize, however, that these were not

[3] One might still maintain that neither the impetus, nor the Galilean, nor the Keplerian, nor the Cartesian views were, or ever could have been, developed in detail, and therefore that—far from being scientific theories—they were not, and never even could have been, anything more than vague suppositions. And from this contention, one might proceed to argue more generally that Newtonian physics was the only *physics*, the only thing that could properly be called a *scientific theory*, that could support Copernicanism. And this would be to reassert the view that the physics called for by Copernicanism was the Newtonian one. Despite the fact that such a view seems to underlie the rather vague interpretations of the historical case given by many scientists, it would be a very hard view to defend. Along with a multitude of other objections, it has the disadvantage of making the expressions "scientific theory" and "correct scientific theory" synonymous.

[4] Nothing is gained in illumination, and much is lost because of highly misleading associations, by referring to such considerations as "aesthetic".

problems, facts requiring explanation, which were recognized as existing with regard to Ptolemaic theories; they were not, and certainly did not need to be, so recognized as long as astronomical theories were considered as mere collections of devices, to be applied to different cases in different ways according to need, for the sake of prediction. In the absence of a realistic interpretation, that is, the existence of these disunities did not constitute a "crisis" for Ptolemaic astronomical theory.) Such greater synthetic unit on the part of Copernicanism was of a high order: all, or most, good theories achieve some degree of such unification relative to past ones, but the synthetic unity achieved by the Copernican theory as compared to the preceding state of things was certainly unusually high.[5]

Let us, then, bring together the positive results of this discussion. We find at least two respects in which the Copernican theory could be called "revolutionary" in a scientific (rather than a cultural) sense.

1. It achieved a very high degree of unification, when compared to its predecessors.

2. Largely because of this synthetic unity, it was easier to view the theory as a "realistic" one: that is, as being representative of the structure of the solar system rather than as a predictive though fictional device. And once this "realistic" interpretation was given, it made sense, indeed was necessary, to reject Aristotelian physics (or portions of it). This did not mean that a "new" physics had to be sought, much less that there emerged an inevitable dialectic leading from Copernican astronomy to the triumph of Newtonian physics. The realistic interpretation of the Copernican theory constituted only one element in that whole story, that total revolutionary transition; but it was an element, one step, and a beginning and therefore an important one.[6]

[5] I am of course fully aware that no precise analysis is available of the notion of "synthetic unity", or even of "relative degree of synthetic unity". However, philosophers err in the most fundamental way in supposing that such judgments as that made here about the degree of synthetic unity achieved by Copernicanism cannot be made without such a prior analysis. On the contrary, *any such precise analysis must be made on the basis of close study of paradigm cases of what are naturally called synthetic unifications.* The general notion of "synthetic unity" may not have been given a precise analysis; but the respects in which the present case achieves such unification, relative to its predecessors, *are* clear, and the adequacy of any analysis of the general notion must be judged in the light of this and other such case studies. In this respect, and to this extent (among others), it may be said that philosophy of science is an empirical discipline.

[6] I am not claiming that "realism" in the interpretation of a scientific theory or concept is a necessary condition of scientific progress. On the contrary, there are many cases in the history of science in which, although, at least initially, an "instrumentalist" or "fictionalist" interpretation of a theory or concept was given, scientific progress nevertheless resulted. It is one of the tasks of the philosopher of science to analyze the circumstances and respects in which "realistic" and "instrumentalist" interpretations can lead to (or hinder) scientific progress. For a study of some such circumstances, see my "Notes toward a post-positivistic interpretation of science", in P. Achinstein and S. Barker (eds.), *The Legacy of Logical Positivism*, pp. 115–60, Baltimore, Johns Hopkins University Press, 1969.

If these are the respects which encourage us to speak of a Copernican Revolution in a scientific respect, then in the case of *this* scientific revolution, at least, there was no radical discontinuity with past traditions. There was change, to be sure, but not of a kind that (like the innovations of Faraday and Planck) could be spoken of as an introduction of radically "new concepts". Nor, I think, can it be claimed that, in this case at least, "revolutionary" reasoning in science is different in any radical way from "normal" scientific reasoning: improved synthesis is something that is achieved, to a greater or lesser degree, by many scientific theories, and—especially once the theories are taken as "realistic" (which they often are not, at first)—most theories, even those more humdrum and ordinary than "revolutionary" ones, raise new problems, calling for new explanations. That is, in the Copernican case—at least with respect to the two features noted here—the difference between scientific revolutions and the reasoning involved in other, more commonplace scientific episodes is a matter of degree, not of kind, and the distinction between pre-Copernican and Copernican theories is not marked by a sharp, unbridgeable break between incomparable ("incommensurable") traditions. Whether similar conclusions can be drawn about all cases of what are or can reasonably claim to be called "scientific revolutions" (for example, whether something more radical, and if so exactly what, goes on in cases like the introduction of "new concepts" such as "field", "expanding space", or "quanta") is, of course, a question that can be answered only by detailed investigation.

On Normal and Extraordinary Science

THOMAS B. SETTLE

Polytechnic Institute of New York, Brooklyn, N.Y.

THE members of this panel were asked to address themselves to some facet of the general problem: was the Copernican Revolution an archetypal scientific revolution? Here I have turned the question around in order to suggest what appears to me to be a more appropriate way to conceive such an archetype, more appropriate, that is, than the way usually assumed by Thomas Kuhn and those who have commented on his *The Structure of Scientific Revolutions*.[1]

It is clear that a great many historians eagerly welcomed Thomas Kuhn's project of developing a general theory of scientific advance.[2] The history of science was (and still is) defining itself as a discipline, and we needed a critical apparatus, a comprehensive set of tools and concepts, that could help us both distinguish interesting avenues of inquiry and get on with the understanding of the historical dynamics internal to science. And we could expect in Kuhn's work a picture solidly grounded in biographical and historical records rather than in philosophical prescriptions. And it is equally clear that from the beginning there have been doubts about Kuhn's ideas. One such doubt, indeed one most forcefully voiced, resulted from the difficulty in arriving at a suitable or respectable notion of the *paradigm*. In response Kuhn him-

[1] First ed., 1962; Second ed., inc. "Postscript—1969", 1970: University of Chicago Press. For further by Kuhn see: *Criticism and the Growth of Knowledge*, Imre Lakatos and Alan Musgrave, eds., Cambridge University Press, 1970. On Kuhn see the other papers in Lakatos and Musgrave; these also cite an extensive literature. But see especially, Dudley Shapere, "Meaning and scientific change", in *Mind and Cosmos* (vol. 3, University of Pittsburgh Series in the Philosophy of Science), pp. 41–85, R. G. Colodny, ed., University of Pittsburgh Press, 1966.

[2] Kuhn's interests were already evident in his *The Copernican Revolution*.

self has refined the concept and given us a model with considerably improved definition and clarity.

Other doubts have remained, however, doubts of a more historical nature. Typically, even sympathetic historians of science have commented that Kuhn's model does not seem to work in the areas of their own special expertise, though it might well be applicable elsewhere.[3] Now if only one person had said this, or if only scholars in one localized speciality had said it, the implied criticism would not necessarily be terribly damaging. But when such remarks are summed across most of the discipline, the criticism becomes serious indeed. We could not long support a very general thesis about the operation of certain historical processes if experts could find no satisfactory instance that exemplified the thesis in detail. If we assume that whatever else the Kuhnian thesis is (an epistemological thesis, for example, or a methodological program), it is first and foremost an historical thesis, then we ought to find ways of retrieving it in our daily rounds as work-a-day historians, or we ought to put it aside and construct an alternative. What follows, then, is an attempt in sketch form to scrutinize one of the sets of Kuhn's assumptions and, in the light of a particular example, suggest some revisions.

One of the distinguishing features of the thesis is the crucial distinction Kuhn makes between two types of science: normal and extraordinary or revolutionary. The general considerations that seem to have engendered and justified the distinction are familiar enough (for example, the observation that most scientists are "journeymen" doing normal research, and not "masters"), and they need not be reviewed in detail. For our purposes it is sufficient to note that Kuhn and his commentators all seem to assume that the distinction is a real one, i.e. that a particular piece of research can be put definitively into one of two pigeon holes, labeled either *normal* or *revolutionary*, and that even the work of one individual can be so divided. Moreover, in addition to the assumed qualitative distinction *per se*, there is a further assumption regarding a definite temporal sequence or alternation of the types. The sequence might be rendered thus:

> . . ., a paradigm or exemplar comes to dominate a field;
> then follows a period of normal science;
> then anomalies arise and precipitate a crisis;
> then follows a period of revolutionary science, ending in the dominance of a new paradigm;
> then . . .;

So the thesis as it is now framed clearly requires the existence of two types of science, and the assertion of a temporal alternation between the two reinforces and hardens the point. We might well ask, however, if the distinction is a real one. And if not, how could the thesis be reformulated?

[3] Such comments were found occasionally in the reviews, but I found them most often in conversation with colleagues.

Everyone is familiar to some degree with the mature science of natural motion Galileo presented in his *Discourses on Two New Sciences* (1638) towards the end of his life. Most will also be aware that Galileo authored a manuscript treatise on motion, never published during his own lifetime but dating from the very beginning of his career in about 1590, in which he expounded a very different theory of natural motion.[4] Since there are some obvious and striking contrasts between these two theories of motion, we could suppose that there was a revolution of some sort, sometime in the transition between the two. What can we learn about this revolution?

One of the central features of Galileo's final and mature theory of motion was his law of free fall, the principle that states that in free, unimpeded fall a body will move in such a way that its distances from rest will be in direct proportion to the squares of the time intervals from rest. As we know, Galileo was in command of this principle a long time before the publication of the *Discorsi*; in 1604, 34 years previously, he mentioned it in a famous letter to his Venetian friend Fra Paolo Sarpi. In 1604, then, Galileo knew that bodies in free fall were continuously accelerated, that in principle this acceleration had no natural closure (i.e. that in the absence of impediments a body would go on accelerating—the implications of this point will be apparent in a moment), and that in such acceleration the distances are as the squares of the times.

What about the earlier theory of motion? For Galileo in the manuscript *De Motu* the important feature of downward motion was that it was *natural* for a body to acquire a uniform motion in fall, a uniform motion proportional to its specific gravity. This uniform speed was not what we would call today a terminal velocity, a velocity determined in part by the properties of the medium in which the body is moving. For Galileo this uniform speed was a property of bodies which would be exhibited, indeed exhibited most perfectly, in a vacuum. The contrast between the two theories could not be more great: the later one seeing natural fall as a continuous unbounded acceleration, the earlier one seeing such fall as a fixed uniform speed.

Having said this, it is fair to say that at either end of this transition there were disclaimers having to do with the difficulties of finding examples of the respective motions in nature. In the final theory Galileo recognized that the accelerated motion would only be exhibited where a body's specific gravity was substantially larger than that of the medium and until the body's motion was impeded by the resistance of that

[4] *De Motu*, translated and published in I. E. Drabkin and S. Drake, *Galileo Galilei: On Motion and on Mechanics*, University of Wisconsin Press, 1960. Much of the supporting evidence for what follows can be found in T. B. Settle, "Galileo's use of experiment as a tool of investigation", in E. McMullin, ed., *Galileo: Man of Science*, New York, 1967. Further considerations are in T. B. Settle, *Galilean Science: Essays in the Mechanics and Dynamics of the "Discorsi"*, Dissertation, Cornell University, 1966. Further important work on the early Galileo has been done by Raymond Fredette, *Les "Des Motu" 'plus anciens' de Galileo Galilei: Prolongomenes*, Dissertation, Institut d'Etudes, Medièvales, Faculté de Philosophie, Université de Montreal, 1969, and by Winifred Lovell Wisan *Galileo's "De Motu Locali"*, Dissertation, University of Florida, 1972. Some of Fredette's results appear in his "Galileo's *De Motu Antiquiora*, *Physis*, anno XIV, 322–48 (1972).

medium. In other words, the essential feature of natural motion was that it was accelerated in a certain way; in practice bodies might fall with uniform motion, but this would be an accidental feature of that motion. In the early work, on the other hand, a uniform motion was its essential feature. Galileo recognized that when a ball was released from rest it did not jump instantaneously from rest to its natural speed; in fact it went through a period of acceleration; but this acceleration was an accidental feature of the motion, accidental in the strong sense that nothing interesting or intelligible could be said about it; it could not be reduced to rational formulation.

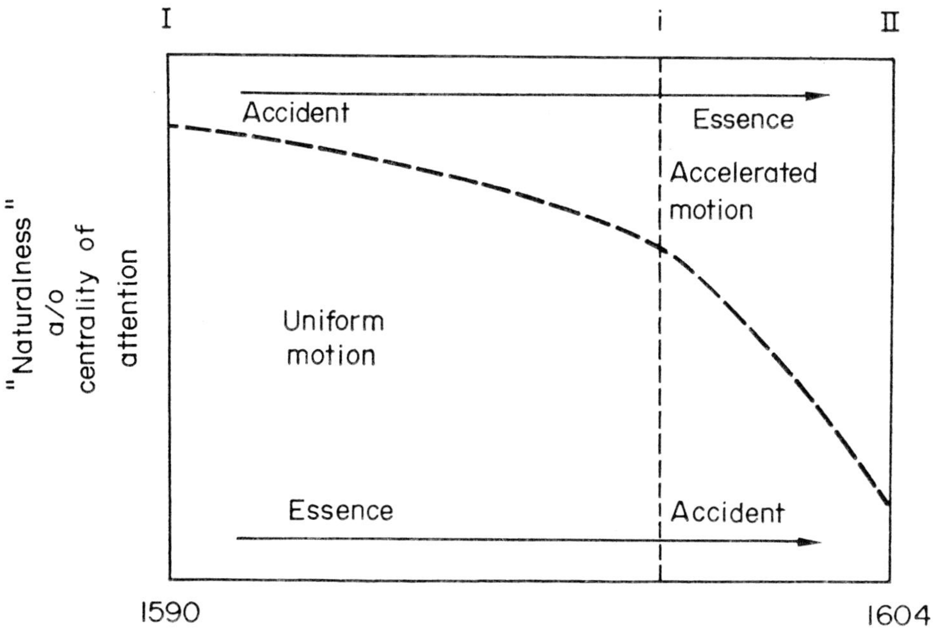

FIG. 53. The problem of "rational formulation".

(See Fig. 53.) In 1590, then, uniform motion occupies stage center. It is the paradigm or principle on which Galileo bases his research, both theoretical and experimental. To use Kuhn's terminology, it is the exemplar, hypothetical perhaps, that sets the tone and direction of Galileo's work. By 1604, there has been a gestalt switch. Acceleration occupies stage center and the law of free fall (along with other paradigms) sets the tone and direction of Galileo's work all the way until the publication of the *Discorsi*.

But how do we get from 1590 to 1604, from point I to point II? I wish I could report that my sense of what went on in this interval conforms to what is usually taken as Kuhn's model: that for a period Galileo pursued his "uniform motion" model, that there followed a period of crisis, and that there was a revolutionary gestalt switch, followed by a new period of normal science. I say, I wish I could do this, but for me it just does not hang that way. On the contrary, in a very real sense there was a strong continuity between 1590 and 1604. It would take too much space to try to defend completely this assertion here, but let me indicate what I mean by drawing attention to one of the transition stages between the two extremes I have diagrammed.[5] While Galileo was elaborating the early manuscript *De Motu* acceleration became more and more of a problem for him. Then at one point, though continuing to hold to the uniform motion model, he came to the decision that even though acceleration was an accident, it might nevertheless be intelligible. He proposed that we could design an acceleration according to which a body's velocity would asymptotically approach the uniform motion required by the primary model. If, for instance, after a certain distance a body had acquired a speed V and then, after the next equal distance it had acquired an additional $(\frac{1}{3})V$, and then after a further equal distance an additional $(\frac{1}{3})$ $(\frac{1}{3})V$, and so on, it is clear that this accelerated motion would asymptotically approach a uniform motion. And if this indeed were the way acceleration occurred, we would be on the verge of describing it in mathematical terms. What are interesting here, I think, are two things: one, the increasing attention Galileo gave to acceleration *within* the context of pursuing his "normal research" into facets of his uniform motion theory; and two, the fact that Galileo has proposed representing acceleration by a defined mathematical series: 1, $\frac{1}{3}$, $\frac{1}{3}^2$, and so on. By 1604, Galileo would propose that bodies accelerate such that in equal time intervals they travel distances corresponding to the series of odd numbers: 1, 3, 5, 7, . . .; this series, of course, is the equivalent of the $S :: T^2$ proportionality, the law of free fall.

What this and other things suggest to me is that there is no useful way to interpret this phase of Galileo's work as one in which there was a sequence of normal science and then a crisis and revolution. What we have, right from the beginning, are the seeds of the final configuration (the seeds, not the mature entity). Acceleration is present from the beginning but playing only a minor role. In the process of working out his ideas, Galileo comes to give it more and more attention. Finally acceleration displaces uniform motion at prime focus. At no point in the process does it seem to make much sense to say that Galileo is now doing purely normal research, or that later he is doing purely revolutionary work. In other words the two, even if they can be distinguished analytically, interpenetrate completely. Nor can it even be said of a particular increment of the total episode that it is purely normal or purely revolutionary; each increment looks both forwards and backwards. What we can say usefully, once we have defined the end points of a particular episode is something about

[5] Despite a flurry of recent work on this period (see note 4), there is much more to be done before any kind of a consensus can hope to be reached. Stillman Drake has recently found impressive evidence for Galilean experimentation in problems of motion, at least in the first decade of the seventeenth century; see "Galileo's experimental confirmation of horizontal inertia: Unpublished manuscripts (Galileo Gleanings XXII)", *Isis*, **64**, 291–305 (1973).

the trend of events between those two points, the set of the tide. If the end points in this case can be characterized as paradigms, the transformation between the two can be likened to a metamorphosis, one in which there appears by the surface of things to have been a radical, discontinuous change, but one in which, when viewed in detail, there were stronger, more significant continuities.

If what I have said holds up, it would seem to entail abandoning the notions of normal and extraordinary science. Not necessarily. At least not in so far as we are still faced with two tasks: one, instructing ourselves about the details of the transition from point I to point II, i.e. doing our research; and two, organizing and describing our results to our colleagues and students. Anyone who has tried to describe accurately the motion of a projectile in purely verbal terms, without using or even referring to a set of Cartesian coordinates or to an analysis of the motion into its component parts, knows how difficult it is. And he will know viscerally the advantages conferred by being able to define a field and to analyze a complex motion (projectile motion, say) into its components (uniform and accelerated, say). Likewise here, it could be immensely useful to be able to describe the situation at each point in the transition from I to II (point "i", say) as having a component of normal science and a component of revolutionary science. In the example cited, for instance, the proposed asymptotic acceleration is neither exclusively normal nor exclusively extraordinary. Rather it is both at the same time. It is normal in so far as it is a result of Galileo's continuing efforts to hold onto his original paradigm or hypothetical exemplar in the face of mounting difficulties in handling acceleration. And it is extraordinary in the sense that, in retrospect at least, it is clearly an important conceptual and mathematical midpoint between I and II. So in analysis we can treat it as having two dimensions, two components, even though we know that in real historical time the two components comprised, in fact, one event or stage in the total episode.

And there are ways, I believe, in which the point of view I have been advocating could have interesting consequences in our research, but a full discussion would lead us too far afield for the present. Let me say only that I expect that if we could allow the historian and biographer to regard normal and extraordinary science as analytical components of some composite reality, instead of really distinct entities in a time sequence, they could gain a useful research tool without being forced to do violence to their subjects. I am suggesting, in short, that we consider the motions of normal and extraordinary science, not as phases in scientific development which occur one after the other, but as components of an ongoing scientific change, of use especially in retrospect in coming to some understanding both of those past events and of the process of change themselves.

Now this is a Copernican commemorative conference, and thus far I have said precious little about Copernicus or his revolution. In fact I had intended to reflect these thoughts and remarks against the career of Copernicus rather than that of Galileo. But as I began to identify appropriate examples or materials I realized that

I would do best by sticking to what I knew of Galileo and only then offering a comment and posing a question to the true Copernican scholars. First, it strikes me that in many treatments of Copernicus it is regarded as paradoxical and somehow improper that a man should exhibit so many conservative characteristics and at the same time produce such revolutionary work. As some medieval theorist of motion might have argued; it is impossible that a man should exhibit two contrary characteristics at the same time. But perhaps, in the light of my remarks, we ought to regard such situations as natural or "normal" in a larger sense, and not paradoxical at all. On the other hand, perhaps the supposed paradoxicality of the situation was never more than a literary pose anyway.

Finally, I wonder whether in discussing Copernicus we ought not to talk of his revolution*s*, plural. After all, Copernicus was a student of astronomy for some 50 years. At the beginning, presumably, he was taught some version of the Ptolemaic system. And over the years, in a series of steps (and not in one great leap), he worked his way to his final position. Each of these steps was a revolution of a sort. One of the early ones resulted in the realization that the Earth was not necessarily absolutely fixed at the center of things. And there were later ones; we know, for instance, that there are substantive differences between the planetary theory presented in the *Commentariolus* and that of the *De revolutionibus*. Each one of these would constitute a "sub-revolution", and many of these sub-revolutions would be related to the complete Copernican revolution in the same way as the Galilean episode I used above is related to the complete Galilean revolution in dynamics. My question, then, is: regarding any one of these Copernican episodes, are there the comprehensive materials and sources still remaining which might allow us to put my conjectures about normal and revolutionary science to the test?

Einstein on Scientific Revolutions†

MARTIN J. KLEIN

Yale University, New Haven

ALBERT EINSTEIN disliked too much talk about scientific revolutions. "The reader gets the impression", he once commented "that every five minutes there is a revolution in science, somewhat like the *coups d'état* in some of the smaller unstable republics."[1] He thought that overusing the term scientific revolution produced a false image of the process by which science develops, and that this was really "a process of development to which the best brains of successive generations add by untiring labor", a process which "slowly leads to a deeper conception of the laws of nature". Einstein did occasionally refer to scientific revolutions but only when the transformations involved were, so to speak, on the scale of the French or Russian Revolutions. I think it is worth our while to try to understand why Einstein, with his unquestionable credentials as a revolutionary, preferred to use the term in a particular, sparing way. Any theory of scientific change that gives revolutions a central and even an exclusive role will have to confront his dissenting opinion.

Einstein's specific remarks rejecting the language of revolution which I have quoted cannot be dismissed as just an overstated response to an individual situation.

† This paper was written at the Institute for Advanced Study in Princeton, New Jersey, whose hospitality and support I gratefully acknowledge. I am also glad to acknowledge financial support at the Institute from the National Endowment for the Humanities under Grant H5426 and at Yale from the National Science Foundation.

[1] Albert Einstein, Statement for the press on Erwin Schrödinger's public announcement concerning his paper "The Final Affine Field Laws". Einstein's statement is dated February 6–7, 1947 and was translated by Ernst G. Straus. The manuscript is in the Einstein Archive. I am grateful to the Trustees of the Estate of Albert Einstein for permission to quote it here.

It is true that he was reacting strongly against a sweeping claim by a colleague to have revolutionized physical thought, but Einstein's statement was completely consistent with the view of the history of physics he often expressed in his later writings. Historians of science are, however, properly cautious when it comes to dealing with general writings by physicists about physics. They are, in fact, inclined to take the advice Einstein once offered. "If you want to find anything out from the theoretical physicists about the methods they use", he wrote, "I advise you to stick closely to one principle: don't listen to their words, fix your attention on their deeds."[2] I intend to follow that advice and look at Einstein's deeds, that is, at what he was actually doing in some of his early work. But I shall come back to his words too, comparing his theory and practice of revolutionary science.

The first important work Einstein did in physics appeared in a series of three papers on statistical mechanics in the years 1902 to 1904. These papers, hardly noticed at the time and generally unknown today, show a number of the features that characterize Einstein as a scientist. Most striking, of course, is his ability. At the age of 23, largely self-taught in physics, having read Ludwig Boltzmann's *Lectures on Gas Theory* but apparently none of his papers, Einstein in effect created the subject of statistical mechanics for himself. As a matter of fact, Boltzmann had never brought out the essentials, the structure, of the statistical theory implicit in his voluminous writings, so sharply and clearly as Einstein did in these early papers.[3] That Einstein obtained the principal features of the theory—the canonical distribution, the equipartition of energy and the physical interpretations of entropy and temperature—is less interesting in the present context than some other aspects of these works.

In the 1902 paper,[4] the first of the series, entitled "The Kinetic Theory of Thermal Equilibrium and the Second Law of Thermodynamics", one can see Einstein's real starting point as a physicist. His self-education had included the books of Helmholtz, Kirchhoff, Maxwell, and Hertz among others,[5] and he had evidently fully absorbed the mechanical world view as it was conceived and employed in the late nineteenth century. His concern in the paper was to complete the *mechanical* basis of thermodynamics, "to derive the second law using only the mechanical equations and the theory of probability". The first section of the paper is called "The mechanical picture of a physical system", and Einstein speaks of representing an arbitrary *physical* system by a *mechanical* system. This is the language of the 1880s and 1890s, the language of dynamical analogies and models used by physicists as varied as

[2] A. Einstein, "On the method of theoretical physics", reprinted in his *Ideas and Opinions*, trans. S. Bargmann, p. 270, New York, 1954.

[3] See my paper, "The development of Boltzmann's statistical ideas", *Acta Physica Austriaca*, Suppl. 10, 53–106 (1973).

[4] A. Einstein, "Kinetische Theorie des Wärmegleichgewichtes und des zweiten Hauptsatzes der Thermodynamik", *Annalen der Physik*, 9, 417–33 (1902).

[5] See, for example, Philipp Frank, *Einstein, his Life and Times*, trans. E. Rosen, p. 20, New York, 1947.

Boltzmann, J. J. Thomson, Helmholtz, and Hertz, all deeply influenced by James Clerk Maxwell.[6] (It is also the language that Josiah Willard Gibbs used in his book on statistical mechanics which appeared in 1902 but was not read by Einstein until some years later.) Einstein was very explicit in his conclusion: "The second law therefore appears as a necessary consequence of the mechanical world picture."

Even in this first article, however, there are suggestions that Einstein was conscious of the limitations of that mechanical world picture and was eager to free his argument from some of its restrictions. At several key points he draws the reader's attention to how little use has been made of the specific form of the mechanical equations. He also points out that the forms of the expressions he has derived for entropy and temperature are much more general than their derivations might lead one to expect. These suggestions for generalization are elaborated in the new analysis presented in the 1903 paper, "A Theory of the Foundations of Thermodynamics".[7] Einstein dropped the word "kinetic" from his title as a sign of the greater generality of his assumptions. The new theory retains only the bare essentials of the mechanical representation of the system—first-order linear differential equations, a unique energy integral and the analogue of Liouville's theorem. There is no need for the particular form of Hamilton's equations, or for the division of the basic variables into the two categories, coordinates and momenta, or even for the separation of the energy into kinetic and potential terms. Einstein saw this generalization as an elimination of the force concept from the theory, a goal he had perhaps derived from his study of Hertz's *Mechanics*.

One can characterize the development of Einstein's ideas on statistical mechanics by saying that the mechanics gradually become less prominent while the purely statistical features of the theory take on more importance. Although it would be inappropriate to discuss the details here, I must at least mention that the 1904 paper[8] contains the first serious treatment of statistical fluctuations ever given.[9] That is to say, whereas Gibbs and Boltzmann had calculated how large energy fluctuations would be and then dismissed them as reassuringly unobservable, Einstein immediately began to search for some system whose fluctuations could be measured. Since the scale of fluctuation phenomena is set by the same constant that determines the scale of molecular magnitudes, as Einstein proved, fluctuation measurements could serve to fix a fundamental natural constant. His search proved successful in 1905 when he showed that colloidal particles, even those large enough to be observed through a microscope, must carry out a completely chaotic perpetual motion,

[6] See M. J. Klein, "Mechanical explanation at the end of the nineteenth century", *Centaurus*, 17, 58–82 (1972).

[7] A. Einstein, "Eine Theorie der Grundlagen der Thermodynamik", *Ann. d. Phys.* **11**, 170–87 (1903).

[8] A. Einstein, "Zur allgemeinen molekularen Theorie der Wärme", *Ann. d. Phys.* **14**, 354–62 (1904).

[9] See M. J. Klein, "Thermodynamics in Einstein's thought," *Science*, **157**, 509–16 (1967).

as a consequence of the thermal motion of the molecules making up the liquid in which they are suspended.[10] If such a random motion, obeying the laws Einstein derived for its mean behavior, did not exist, one could well doubt the whole structure of statistical mechanics. Its real test had to come in a fluctuation phenomenon, since the averages it predicted necessarily coincided with the laws of thermodynamics. And conversely, as Einstein wrote, if his predicted motion were observed, "then classical thermodynamics could no longer be looked upon as precisely applicable even to bodies of dimensions distinguishable in a microscope", and the scale of molecular magnitudes could be determined. The motion Einstein invented had in fact been known for close to a century as the Brownian motion.

Now we might wonder whether this fluctuation theory—which Einstein and the great Polish physicist Marian von Smoluchowski developed in parallel during the next few years—ought to be designated as revolutionary. Certainly many of their contemporaries considered it to be so. Its experimental confirmation by Jean Perrin, The Svedberg, and others is usually considered a definitive proof of the ancient hypothesis that matter is atomic in structure.[11] Even many of those who needed no further convincing about the reality of atoms were startled to learn that the great second law of thermodynamics had its limits of validity, that the irreversibility the second law demands of all natural processes is only an apparent irreversibility: the same statistical theory that assures us of the virtual certainty of irreversibility in the macroscopic world also requires the rapid and frequent recurrence of states for sufficiently small systems.

Einstein, who fully recognized the significance of his work, did not consider it revolutionary. How could it be when it was only a fully consequent working out of the implications of the mechanical world view? As for thermodynamics, he did not see it as undermined by having its validity confined to certain limits; it was not being discarded or replaced. Again, how could it be? Thermodynamics, for Einstein, was a theory of an essentially different kind, a theory of principle, to use his name, rather than a constructive theory that said something about the nature of the world, like statistical mechanics or any of the other theories that explained phenomena in mechanical terms.[12] Thermodynamics does not attempt to explain in detail what the world is like. It starts from the fact that perpetual motion never occurs in normal experience and asks the question: "What must the laws of Nature be like so that it is impossible to construct a perpetual motion machine?" The answer is a theory that relates a variety of macroscopic properties to each other but gives no structural

[10] A. Einstein, "Die von der molekularkinetischen Theorie der Wärme geforderte Bewegung von in ruhenden Flüssigkeiten suspendierten Teilchen", *Ann. d. Phys.* 17, 549–60 (1905).

[11] Jean Perrin, *Les Atomes*, Paris, 1913. Also see his report to the first Solvay congress, "Rapport sur les preuves de la réalité moléculaire", in P. Langevin and L. de Broglie, eds., *La Théorie du rayonnement et les quanta*, Paris, 1912.

[12] A. Einstein, *Out of My Later Years*, pp. 54–55, 77, New York, 1950.

information about the systems to which it is applied. That is why Einstein could write about thermodynamics in his "Autobiographical Notes": "It is the only physical theory of universal content concerning which I am convinced that, within the framework of the applicability of its basic concepts, it will never be overthrown."[13]

Just because the young Einstein was so deeply impressed by the mechanical world view, by the Newtonian program of understanding all natural phenomena as consequences of the various kinds of forces between material particles, he did find one nineteenth-century development revolutionary. It was a development that offered the possibility of an alternative to this world view, one that might require putting physics on a thoroughly new foundation. "The most fascinating subject at the time that I was a student", Einstein wrote in that same autobiography, "was Maxwell's theory. What made this theory appear revolutionary was the transition from forces at a distance to fields as fundamental variables."[14] Maxwell, to be sure, had conceived of Faraday's field as a dynamical system and was convinced that its detailed mechanical structure would eventually be determined. His immediate successors shared this conviction, but it had faded before the end of the century as repeated attempts to determine the field's mechanism proved fruitless. The field itself became the basic conceptual structure of the theory; and physicists no longer sought to ground it in mechanical principles. It was enough to deal with this set of continuous functions, defined throughout space, which was determined by Maxwell's partial differential equations.

It even seemed for a time as though electromagnetism might replace mechanics as the fundamental science, that in terms of which all phenomena were to be understood. At the turn of the century a number of physicists were trying to explain mass as due to the inertia of the electromagnetic field, and some had visions of creating an electromagnetic world view to replace the mechanical one which had just proved inadequate.[15] Things did not work out so simply. As Einstein put it, this "revolution, begun by the introduction of the field, was by no means finished".[16] The electromagnetic field theory was incapable of accounting for material particles and their masses. The most successful theory of the time was H. A. Lorentz's theory of electrons which needed and used both mechanical and electromagnetic concepts on an equal footing. Lorentz's theory employed particles and fields satisfying, respectively, Newton's and Maxwell's equations, and clearly separated their essential roles. The very clarity of the electron theory (as compared to Maxwell's original

[13] A. Einstein, "Autobiographical notes", in P. A. Schilpp, ed., *Albert Einstein: Philosopher-Scientist*, pp. 1–95, Evanston, Illinois, 1949. Quotation from p. 33.

[14] *Ibid.*, p. 33.

[15] See Russell McCormmach, "H. A. Lorentz and the electromagnetic view of nature", *Isis*, **61**, 459–97 (1970); "Einstein, Lorentz, and the electron theory", *Historical Studies in the Physical Sciences*, ed. R. McCormmach, **2**, 41–87 (1970).

[16] A. Einstein, *op. cit.*, note 13, p. 37.

formulation) served to emphasize what Einstein called the "disturbing dualism"[17] in the foundations of physics.

In the essays from which I have been quoting, Einstein stressed the importance of this incomplete revolution for his own development as a physicist. He had made the problem posed by the "disturbing dualism" his own problem. His efforts to unify the foundations of physics and thereby to achieve the revolution begun by Faraday and Maxwell occupied his life. Einstein's retrospective accounts of this attitude toward his work are completely consistent with a direct reading of his early papers dealing with this central problem. Let us look at some features of these for further insights into Einstein's attitude on what constitutes a scientific revolution.

His 1905 paper on Brownian motion came between two much more momentous works—the papers that introduced the light quantum hypothesis[18] and the special theory of relativity[19] into physics. Both grew out of Einstein's concern with that disturbing dualism in the foundations of physics. Both demonstrated that this concern was not primarily due to a desire for a more beautiful theory or for greater logical economy; it came, rather, from Einstein's recognition that some of the gravest unsolved problems in physics had their origin in the incompatibility of the two disparate fundamental theories—mechanics and electromagnetism—when they had to be brought to bear together. Both the hypothesis of light quanta and the special theory were his partial responses to those problems of incompatibility.

Einstein described the hypothesis of light quanta as a "heuristic viewpoint" that might be valuable in trying to understand the interaction between radiation and matter. Fully recognizing the audacity of his suggestion (he referred to it as "very revolutionary" in a letter to a friend[20] in 1905), Einstein proposed that light be considered as a collection of independent particles of energy, behaving much like the material particles of a gas.[21] He was proposing that physicists consider abandoning the electromagnetic-wave theory of light, the greatest triumph of Maxwell's field theory and of all nineteenth-century physics. Einstein was brought to this extreme step by a simple argument about the Brownian motion of a charged material oscillator, just such an oscillator as Max Planck had used in his recent studies of black-body radiation. Requiring that the oscillator be in thermal equilibrium with a gas—obeying statistical mechanics—and with radiation—obeying Maxwell's equations—led to an intolerable result: the theory could not predict the frequency

[17] A. Einstein, *op. cit.*, p. 37.

[18] A. Einstein, "Über einen die Erzeugung und Verwandlung des Lichtes betreffenden heuristischen Gesichtspunkt", *Ann. d. Phys.*, 17, 132–48 (1905).

[19] A. Einstein, "Zur Elektrodynamik bewegter Körper", *Ann. d. Phys.* 17, 891–921 (1905).

[20] A. Einstein to C. Habicht, n.d., Spring 1905. Quoted in Carl Seelig, *Albert Einstein, A Documentary Biography*, trans. M. Savill, p. 74, London, 1956.

[21] See M. J. Klein, "Einstein's first paper on quanta", *The Natural Philosopher*, ed. D. E. Gershenson and D. A. Greenberg, 2, 57–86 (1963).

distribution of the radiation at equilibrium. Faced with this failure of the basic constructive theories Einstein turned to a theory of principle, thermodynamics, and to the Boltzmann principle, to which he assigned the same status. With their help he showed that the empirical form of the distribution at high frequencies suggested the light-quantum hypothesis, a hypothesis that he immediately used to account for a variety of perplexing phenomena including the photoelectric effect. Einstein emphasized, however, that the light quantum was only a hypothesis, a working suggestion, that might be a heuristic guide in constructing the new theory that was needed. He did not have a theory of light quanta.

In his paper "On the Electrodynamics of Moving Bodies" Einstein used a very different approach to deal with another problem arising from the incompatibility between mechanics and field theory. Maxwell's electrodynamics, even in Lorentz's clarified version, assumed the existence of a luminiferous aether. This was originally considered to be the medium in which electromagnetic phenomena took place and, in particular, the medium in which electromagnetic waves propagated with velocity c, independent of the speed of their source. It retained its existence in Lorentz's theory as the uniquely preferred frame of reference. Electromagnetic field theory did not, therefore, obey the classical relativity principle of mechanics, which was already well understood in the seventeenth century. The theory predicted a variety of effects which should make it possible to distinguish between an observer at rest in the aether and another observer moving uniformly through the aether. These effects were never observed. That is to say, electromagnetic phenomena *were* consistent with the principle of relativity, even though the speed of light was independent of the speed of its source, as expected for waves propagating in a medium. Einstein resolved this apparent contradiction in his first theory of relativity. He did it by creating a new theory of principle, on the model of thermodynamics. Just as the latter theory was, essentially, the answer to the question: How must the laws of Nature be restricted so that there cannot be a perpetual motion machine?, so the new theory, special relativity, answered the question: How must the laws of nature be restricted so that there are no privileged observers?[22] The key to Einstein's resolution of the difficulty was the new analysis of simultaneity that he made after recognizing that "time was suspect".[23]

Once again, just as in the light quantum hypothesis, he did not claim to have developed the new fundamental theory that was needed. Special relativity, despite the new understanding of space-time that it introduced, despite the power of its requirement that acceptable physical theories must be invariant under Lorentz transformations, was still only "a heuristic principle", as Einstein wrote in 1907.

[22] A. Einstein, *Lettres à Maurice Solovine*, pp. 18–21, Paris, 1956.

[23] Robert S. Shankland, "Conversations with Albert Einstein", *American Journal of Physics*, **31**, 47–57 (1963).

It was "only . . . a principle that allows one to reduce certain laws to others, analogously to the second law of thermodynamics".[24]

Einstein knew from the outset that his early work had not solved the problem of providing a new unified foundation for physics, and for him that was always *the* problem. The two heuristic principles of 1905 led to two very different kinds of developments—the general theory of relativity and quantum mechanics. But again, neither general relativity, which he created, nor quantum mechanics, to which creation he contributed in a major way, did the one great thing. Einstein went on struggling to construct a unified field theory, year after year, decade after decade, holding to his vision although it was shared by almost none of his colleagues. He could not accept the substitute offered by quantum physics, which he recognized as "the most successful physical theory of our time".[25] As he saw it, quantum physics had abandoned any attempt to give a complete description of physical reality. If a completed field theory could not furnish that description, and could not provide the unified foundations—and he often expressed doubts on this point in his later years[26] —then some wholly new algebraic approach would have to be invented; for the new foundations had to be built.

When Einstein reacted skeptically to claims that this or that new discovery or new theory had revolutionized physics, he did so in the name of the True Revolution. The old regime of the Newtonian world view had been abandoned, but its genuine successor must provide a comprehensive, consistent, and unified picture of physical reality to replace what was gone. Provisional structures of thought that offered less could be valued for what they were, but Einstein refused to grant their claims of a revolution accomplished.

Einstein never restricted his skepticism to other people's ideas; he was always prepared to doubt his own, including those on the history of science. Speaking of his ideas as those of "one who has struggled his life long with all his strength to clarify and improve the foundations of his science", he went on: "The way in which he views the past and present of his subject may depend too strongly on what he expects from the future and what he is striving for in the present. But this becomes the fate of anyone who has intensively worked his way into a particular world of ideas. In this he is like the historian who likewise—even if perhaps unconsciously—groups actual events around ideals suitable to human society, ideals which he has formed for himself."[27]

[24] A. Einstein, "Bemerkung zur Notiz des Herrn P. Ehrenfest: Translation deformierbarer Elektronen und der Flächensatz", *Ann. d. Phys.* 23, 206–8 (1907).

[25] A. Einstein, *op. cit.*, note 13, p. 80.

[26] *Ibid.*, p. 93 footnote. Also A. Einstein, *The Meaning of Relativity*, 5th ed., pp. 165–6, Princeton, 1955.

[27] A. Einstein, *Mein Weltbild*, ed. C. Seelig (reprinted Berlin, 1968), pp. 113–14.

DISCUSSION

EDWARD ROSEN: I should like to start the discussion with a few comments on various aspects of the splendid talks we heard earlier this afternoon. I should like first to refer to what has been called the 80–34 syndrome. Let me remind you very briefly what this illness refers to. It is asserted by certain persons that the claimed greater simplicity of the Copernican astronomy over the Ptolemaic consisted in a reduction of the number of circles, epicycles, epicyclets of the second order, and so on. What is sometimes forgotten in this discussion is that Copernicus advanced such a claim at the end of a juvenile work, namely, the *Commentariolus*, the authorship of which is now dated between the years 1508 and 1514. In other words, three decades before his mature work, in the closing paragraph of his *Commentariolus*, he made the very ecstatic claim that the sum total of thirty-four circles sufficed to explain what he called the "entire ballet of the planets". What is overlooked in the discussion of this 80–34 syndrome is that no such claim is reiterated in the mature work. You may wonder why. The answer is very simple. It is contained in an earlier section of the *Commentariolus*, his early work, in which Copernicus says "I reserve the mathematical demonstrations for my larger work". It is obvious, therefore, that when he wrote the *Commentariolus* he had before him a simplified sketch of the cosmos. You will recall that the famous diagram of the cosmos in chapter 10 of the first book of the *Revolutions* contained only a single circle for each cosmic body. I put it to you, therefore, that when Copernicus wrote the *Commentariolus*, he believed that the sum total of thirty-four constructions sufficed for what we might call the simplified non-mathematical view of his new cosmos. I repeat that he did not reiterate any such claim in his mature work on the *Revolutions*. Hence, when Ernst Zinner and those who have followed him count up cycle after cycle, as though they were reporting the superbowl, 80 against 34, we are now entering the fourth quarter and so on, they are really missing the point completely.

I should like to turn next to a matter which is very often discussed, and that is what Owen Gingerich has emphasized as the aesthetic approach of Copernicus to his astronomical and cosmological problems. This approach is sometimes treated—not by Gingerich I am sure—as though the idea of the harmony of the Universe were something new, conceived for the first time in the history of mankind by the inventive intellect of Nicolaus Copernicus. Of course, nothing could be farther from the truth. The idea of celestial harmony, as far as the Western world is concerned, has its roots deep in Greek antiquity, not to speak of what went on in the Far East, of which I am profoundly ignorant. Hence, there is of course nothing new or revolutionary in Copernicus' insistence on the harmony in the universe. And this emphasis is to be coupled with his Christian faith in the creation, since he believed in a creative God, who was highly intelligent. Obviously a highly intelligent artisan does not create a disharmonious piece of work.

I wish to call to your particular attention in this connection that Copernicus knew intimately one great work by Ptolemy, which he revered and followed very closely and yet modified in several essential respects, which you may or may not choose to call revolutionary. Frankly, I do not pay much attention to that sort of verbiage. However, there is another work by Ptolemy of very great significance which Copernicus knew nothing about, certainly not at first hand. This later work by Ptolemy, the *Hypotheses of the Planets*, followed the Syntaxis, which is sometimes miscalled the "Almagest". In his *Hypotheses of the Planets* Ptolemy described the harmonious construction of the Universe in accordance with a principle which has been very aptly labeled "The Full Universe". Adopting this conception, let us move from the center of the Universe outward to the nearest planet, Mercury. The greatest distance of Mercury from that center must be exactly equal to the least distance of the next planet, Venus, and so on outward to the ultimate boundary. This is a harmonious universe in which there are no gaps.

Therefore, we must never think of Copernicus as having originated the idea of the harmony of the universe, or as having thought that he was bringing a new element into the picture. What he thought he was doing, was to eliminate the disharmony in an older picture and to restore the true harmony. I think this is the correct historical estimate of Copernicus' attitude toward this great problem of the harmony or disharmony of the cosmos.

Now in this connection it is often pointed out that when Copernicus changed the cosmological status of the Earth from a central body to a planet of the Sun, he thereby diminished the importance of man in the cosmos. I think that from one point of view this conclusion is undoubtedly correct. You will remember that when Sigmund Freud dealt with this problem, he spoke of the three mighty blows that have been delivered to the self-esteem of mankind. Naturally, in his judgment he himself delivered the most important blow, the psychological, by his discovery of the unconscious. But Freud assigned the first such mighty blow to Copernicus, who transferred man's abode from the Earth as the central body to a less important position, namely, that of the third satellite of the Sun. This removal is sometimes regarded as a diminution of man's status. But a very ingenious Copernican, by the name of Johannes Kepler, said: "Oh, no, by moving man away from the center to the third satellite of the Sun, we have enhanced his cosmological status, because God is very smart and we are his beloved creatures." If He had stuck us in one little place in the center, what could we see? We could glimpse very little of His gigantic handiwork. Whereas, if He put us on this earthly boat, which goes around and around the Sun like the boat that goes around Manhattan Island for the tourists, or like the rubberneck bus that goes around lots of cities, that shows you how smart He is. He enhanced our status by putting us on the moving planet, Earth, so that we could admire his entire handiwork. In this way Kepler thought to overcome a pseudo-obstacle to the cosmological innovation of Copernicus.

I have one more remark to make—just by way of suggestion to my good friend, Thomas Settle.

I would like to recall the dates 1590 and 1604. There is a passage in Galileo's *Dialogue* which some people do not take seriously. In it Galileo mentions a mysterious visitor from the north, whom he attaches to the University of Rostock. He said that this man came to Italy and delivered a talk on Copernicanism to the Italian Universities, and that this lecture had an impact on the mind of a certain student. Now I have a feeling, which I cannot prove, that this student's name was Galileo. Nobody has ever been able to identify this visitor from Rostock. Possibly there never was any such visitor, and it may be that Galileo wished to cover up the traces of his conversion to Copernicanism or that he misremembered—I do not know which. But, at any rate in a very famous letter which Galileo wrote to his fellow-Copernican, Johannes Kepler, in the year 1597 he said: "Many years ago I became converted to Copernicanism." What I want to suggest, in connection with the date of this letter, is that in Galileo's essay of 1590 there is no mention of acceleration, or if there is, it is not essential, whereas by 1604 it had become so. At some time between these two years Galileo became a Copernican. I take it, that means he read Copernicus' *Revolutions* and if he did, then he saw that highly pregnant passage concerning the acceleration of a falling body, introduced with that very light touch which Copernicus always uses when he is departing from Aristotelian theories. I offer that as a humble suggestion to those who are specializing in trying to understand how Galileo's thinking changed between his two treatments of the subject of motion.

W. IWANOWSKA: I am neither a historian nor a philosopher, but an astronomer interested in Copernicus mainly because I find myself in the city of Toruń where Copernicus was born 500 years ago. I wish to thank the Organizing Committee most sincerely for its invitation and its help which enabled me to attend the conference, which I have found very interesting and informative (I think Dr. Dobrzycki will join me in extending thanks for his invitation). If you will permit me, I would like to present our view of Copernicus, and I speak in the name of a group of about thirty astronomers.

We have heard many times the words "scientific" and "cultural" revolutions. I would prefer to refer to it as an evolution of modern science beginning with Copernicus and the scientific milestones which followed, which included the names of Galileo, Kepler, Newton, Einstein up to the present day of relativistic astronomy with the concepts of gravitational fields leading to superdense stars like the recently discovered neutron stars and the not yet discovered gravitational collapse in black holes. It is a logical sequence of discoveries—perhaps revolutions—of which each one is a natural and logical consequence of the preceding one. Copernicus, I think, has the exceptional place in this sequence of developments in science in that he made the first step. The first step was not an easy one because, as we have seen, it took 100 and sometimes even 200 years before it was accepted. He helped overcome the limitations of the human instinct and human mind—limitations which could be called an anthropocentrism, where the human race envisioned themselves as being in the center of the world.

This is what I believe many other astronomers think is the primary role of Copernicus in the development of modern science.

BENJAMIN NELSON: In the present setting, some of us might be tempted to forget critical elements of the cultural and religious backgrounds in the response to Copernicus and the struggles over his work. Surely, we need to remember the distress—even dread—that many believers must have associated with the apparent implications of the Copernican hypothesis. A heliocentric universe with annual and diurnal motions of the Earth was not the cosmos as imagined and experienced in Christian teaching and life.

A. O. Lovejoy and others reminded us a long while ago that central Christian notions, such as the doctrines of Atonement and Redemption, underwent very heavy strain wherever the influence of the new science came to be strongly felt and was widely diffused. I would expect that Stillman Drake will make some reference to related horizons when his turn comes and tell us about the sentiments stirred by Bruno's images of an infinite Universe.

My main reason for rising, however, is to speak of another issue, the course of the ongoing discussions—including Thomas S. Kuhn's frequent re-formulations of his own concepts—of "the structures of scientific revolutions". Some day we shall have a history of the many diverse interests that have been at issue in these debates. Perhaps the most interesting aspect of that account will be the extraordinary medley of perspectives and ranges of meaning which were caught up in the developing polemics.

Margaret Masterson, one of the participants in the symposium, *Criticism and the Growth of Knowledge*, edited by Lakatos and Musgrave, at one point lists twenty-one meanings for the Kuhnian use of the word "paradigm". Since that time there have been others who have added meanings that were not clearly focused on that occasion. But the point that is not yet well enough recognized is that latterly—Kuhn notwithstanding—there has been a notable shift in the way Kuhn himself has talked about the elements involved in the structures of scientific revolution.

The sense I have is, that Kuhn's itinerary describes a course from a mode of analysis which retained certain kinds of connections with images of logics and structures of rationales to a horizon that is very largely social-psychological in a way —it does not gladden me to say this—that narrows rather than expands the field. Kuhn is now close to having pointed himself into a corner by adopting the idea, that we only have transformations when we have agreements among the members of the community defined very narrowly within highly evolved sciences comprising of sharply defined disciplines. In the end, therefore, true access is largely barred to precisely those developments in the histories of peoples and the struggles among them over definitions—of and access to—knowledge and techniques which lie below the critical expansions and contractions in the communities of discourse, reference, and language.

In making these modifications, Kuhn moves closer all the time to Robert Merton's notion which was already clearly hinted at in the latter's precocious doctoral dissertation completed in 1936, first printed in 1938, and now available with an important new Preface under the title, *Science, Technology and Society in Seventeenth-century England.* In Merton's view, the central developments within sciences are often best understood in terms of "institutional interchanges" of one sort or other. On my reading this restriction of perspective was a critical step on the road that had few turnings and presented increasingly limited horizons.

There was another road. It was not the one that was taken by Merton or Kuhn, and that other road only comes into view when we widen our sights as to the proper objects of our scrutiny and analysis. And when we expand our vision we will find ourselves wanting to read again and again such writers as Max Weber, Emile Durkheim, Joseph Needham, and all those other men—including Sir Henry Sumner Maine—who have actually dealt with the comparative historical sociology of "sciencing" in civilizational perspective across time.

I take the liberty of reporting that I have written in greater detail about those matters in an essay-review I have done of the 1970 reissue of Merton's aforementioned masterwork, his doctoral dissertation. My review appears in the *American Journal of Sociology* for July 1972.

ERNEST NAGEL: I hope Professor Gingerich will clarify one remark he made toward the end of his paper. If I did not mishear him, he declared that the Copernican theory was "founded" not on new data but on "insight". It is possible that he used the word "founded" in a psychological sense, to refer perhaps to one of the causal determinants that led Copernicus and others during the following years to advocate his theory; and if this is what Professor Gingerich meant to say, I do not question his claim.

However, if he used "founded" in a logical sense, to denote the evidential grounds that warrant in some degree acceptance of the theory, then his remark seems to me unfortunate. For in the first place, while "insight" is undoubtedly needed to construct a scientific theory and to assess the evidence for it, I do not believe that insight ever constitutes *evidence* for a theory. Insights are claims about some subject matters, and claims are not self-certifying. Secondly, although the main burden of Professor Gingerich's paper is a forceful criticism (made in the context of an examination of the Copernican Revolution) of some major contentions in Thomas S. Kuhn's influential account of the structure of scientific revolutions, to hold that the Copernican theory was logically based on "insight" is to nullify in considerable measure that criticism. Many readers of Kuhn's book have been persuaded by it that in general a new theory in science is not adopted because there are rational, evidential grounds for it; on the contrary, a theory is allegedly adopted when a scientific community is "converted", without a careful evaluation of evidence, to a new way of looking at things, to a new "insight" into them. The remark of Professor Gingerich on which I am commenting can be construed to give support to this belief. But if such an interpretation of what he said is mistaken, it is desirable that he makes this explicit.

OWEN GINGERICH: I was wondering what Professor Nagel was going to say about that statement. But what I said was: "Copernicus' radial cosmology is founded not on new observations but on insight." What I am claiming is that nothing new had come up immediately before Copernicus that forced him to adopt heliocentrism. It was looking at the old data in a fresh way and it could have been done any century along in there.

I would change the statement to read: "Copernicus' radial cosmology came forth not from new observations but from insight." Does that satisfy you?

EDWARD ROSEN: Professor Gingerich's paper is a most interesting refutation of the myth that there was an astronomical crisis in the early sixteenth century, and I think this paper is convincing as far as it went. I don't know how much farther you would want to take this investigation, which would require a great deal more research in other countries and other writers. But, what I primarily wanted to say was that, in opposition to Thomas Kuhn's idea of a sixteenth-century crisis, one can counterpose the call for a renovation or restoration or renaissance of astronomy which is often heard in the fifteenth century. For example, Regiomontanus says again and again, that Ptolemy exists in a corrupt text, the corruption making the text unsatisfactory for astronomers, who can solve their problems only by means of new observations.

Once one has this framework of the renaissance of astronomy in the fifteenth century, it is possible to fit Copernicus into the historical development in a way that is more consistent with Professor Gingerich's critique. For example, Regiomontanus may be viewed as the restorer of Ptolemy and of ancient astronomy by means of fresh observations and critical revisions of the received text. This work of restoration went as far as it could without that new insight of Copernicus, which Professor Gingerich mentioned. Why is it that nobody opted for the geokinetic system between the death of Regiomontanus in 1476 and the emergence of Copernicus around 1510? The fact that it is Copernicus who actually made the leap to geokineticism tends to support Professor Gingerich's thesis that the Copernican system was basically the result of new insight rather than simply a response to an unsatisfactory Ptolemaic system.

JEROME R. RAVETZ: I would be grateful for a few minutes of your time because this evening I am to be lecturing to what I am told is an open meeting and a general audience; I fear that I will be constrained to say a lot of general things. I have been sitting here listening to these extremely valuable and exciting talks all day, and I would like just a few minutes to make some remarks at that level of the discussion. I want to say a couple of things about the Copernican revolution and then one or two preliminary things about Professor Kuhn's concept of scientific revolution.

The first thing is that the Copernican Revolution was a process that extended over a very long time. Professor Rosen has given us a valuable reminder that there were people doing astronomy just before Copernicus who, in fact, conceived themselves as making a reform or a renaissance or a revolution. These three terms can be used interchangeably. I might remark that I had not known of the claim by Regiomontanus, it fits in very well with the humanist approach to the restoration of classical learning across a variety of fields in the later fifteenth and early sixteenth century. One has a similar sort of thing in medicine, when there was a rediscovery of Galen by scholars in the very late fifteenth century. This involved a rejection of the muddy and confused tradition of Arabic original sources in medical education as it was then, and a demand to re-edit, to translate Galen, and then of course to purify the text of Galen by new observations. And it is interesting that just as Copernicus could be thought of as a perhaps second or third generation of restorers of astronomy who made the great breakthrough, his position is precisely analogous to Vesalius who himself was a second generation of restorers in anatomy and medicine. So, the Copernican revolution is integral to the rise in modern science. I think, we can all agree, that it took a certain shape in the late fifteenth and early sixteenth centuries, and then another shape in the later sixteenth century. It had its origins, its first announcement and then later it became involved in a complicated way in a revolution or a new move in the philosophy of nature. Thus we have the Copernicanism of Galileo and Descartes which was after all very, very different from the Copernicanism of Bruno or even Campanella, with Kepler perhaps as the bridge. There were several major stages, so that the Copernican revolution could be said more or less to start, let us say not long after 1473, if one takes his predecessors and goes on to 1673, or even to 1773, when finally the book was on the point of being removed from the Index. That took 300 years, during which time it was changing continuously; any simplistic characterization of this long event as one thing is going to be a failure. So we must as historians look closely at which phase of this process we are discussing when we try to make philosophical morals.

Now there are (I should say to Professor Nagel) several very deep problems for the philosophy of science in the earlier phases of the Copernican revolution. We do not have a direct statement by Copernicus of reasons for adopting the new cosmology, which we now consider to be fully adequate. There are scattered among the writings of Copernicus several remarks about why we must change our cosmology. There is the Introduction to the Revolutions; there are the seven reasons given by Rheticus, why we must abandon the hypothesis of the ancients. When we look at these we do not find something that *we* can pick up and say: "Right! That is it!" They are all suggestive evidence in one way or another; some of them we now reject. And so we have a severe problem from the point of view of the methodology of science of justifying in our own terms the commitment of Copernicus and his immediate followers, the commitment of these people to the truth of the heliocentric system.

In this connection, I think that the point concerning "Harmony", of which Professor Rosen has reminded us, is not a new or mystical fancy with Copernicus. It was a very sound, old, established principle of scientific or philosophical explanation. I myself believe that this principle of harmony does certainly exist in some aspects of Copernicus' early works and can also be shown in his later work. My evidence for this comes from a paper by K. P. Moesgaard of some years back about a most peculiar time constant which appears in the Revolution, a period of 1717 Egyptian years. This period governs the basic slow motions of the elements of the heavenly system. It is the period either 1717 or 3434 which governs the figure 8 motion of the Earth's axis which explains the (totally spurious) trepidation effect, the (nearly totally spurious) variations of the obliquity of the ecliptic up and down, and also the motion of the Sun in a little circle around the center of the Universe, which helps to sort out slight changes in the length of the year. And with the slow motions of this same period, itself a rather interesting number 1717, Copernicus could have believed that he had finally reduced the world system to harmony—that he then had a consistent definition, not only of the length of the day, which is not quite trivial, but also the length of the year which is highly non-trivial. And in terms of the structure he had set up for the Earth in the third heaven, at long last there was coherence and a subtle numerical harmony. This is arguable: I will not say anything more than that.

There is also the very deep problem that Professor Nelson has raised, on the influence on society and culture on the further progress of the Copernican revolution. Because here I think we are abutting on a very, very difficult, very deep problem. Are we to say, as we look at the further progress of the Copernican theory through all of its transformation, how long was it before there was a decisive proof? It does appear really that the matter was effectively settled when Newton's *Principia* became accepted by all competent practitioners as the only way to go forward in mathematical astronomy. And also such competent practitioners, or whom there might have been a couple of dozen or less, were accepted by the community of the learned in general and because of *that* support, by most of the less educated. So sometime in the eighteenth century, people said: "Well, Newton is right; you cannot do astronomy without Newton's system, and that assumes heliocentricity, so therefore Copernicus is correct." But we have this long intervening period; are we then to say with Professor Kuhn (as caricatured by his critics), that a change of fashion produced a switch between incommensurable paradigms? Or should we understand Professor Nelson as saying that this fundamental transformation in our science was merely the effect of very deep changes in the way educated people looked at the world? Now I hope that it does not reduce to either of these simple versions; if it is then a great deal of the strength of science and reasons for its autonomy are weakened.

As far as I can take it at the moment, it would be to say (and this is not satisfactory) that a hostile culture could strangle a new idea. In other words, if there may have been a Copernicus in China in the year 1300, there was simply nothing there for the seed to grow in and so it died unnoticed. So we may say that a hostile culture can stop truth from developing, but once the truth starts to move, then it necessarily develops as a single and unique truth, dependent on nature and independent of culture. This is at least the beginning of an answer, but I would not wish to say it is anything close to the final answer.

The third point I want to make concerns Professor Kuhn, because I think we can now begin to stop arguing with him and accept him as an historical phenomenon. We all know how the term "paradigm" has had some dozens of uses, how Professor Kuhn's interpretation of the thing develops overtime. Yet we are all stuck with the idea of "paradigm"; we cannot get away from it. We can also say that the very deep insight with which Professor Kuhn began, is about a quarter of a century old. If you look at the preface of his book, it is describing what it was like when he was a research student. I have reread the book quite recently with new eyes, to use it in philosophy of science lectures. What has struck me about it is that in the preface there is an image of normal scientists who are very nasty men. All they want to do is force nature into conceptual boxes, provided by their professional education. When a threatening innovation comes along they try to suppress it, indeed even by administrative means. And it is only when they can no longer evade the challenge, when everything is blowing up in their faces, then they admit that things are really bad and consider alternatives. I am curious to know what gave him this idea of the practice of science. Was this typical of Harvard University in 1947? I do not know, I was not there. I was at Swarthmore, and it was not like that at Swarthmore. It seems, then, that Professor Kuhn saw this rather uninspired and uninspiring practice around him (which he records as the practice of science) and contrasted it to a certain image of "true" scientific growth that he learned from the history of science lectures given by Conant or Nash. And the contrast between the two may well have been most upsetting to him; he just could not put this together. Out of this dichotomy came the two sorts of science—the "normal" science which he saw going on all around him, and a wonderful revolutionary science which is described in history books. From this insight came his theory; and the weaknesses of the theory reflect, I think, the narrowness of the basis of experience which gave rise to it. And it is only because it happened to contain a great deal of truth, and also to be advanced by a man who is very profound and very honest, that the thing has been with us so long. This is my own contribution to the history of a piece of work which has dominated our effort and will doubtless go on a bit more.

CARL B. BOYER: I am a mere mathematician and hence very brief; but inasmuch as we are speaking of scientific revolutions, I should like to call your attention that if there is such a thing as a mathematical revolution, it certainly is quite different from a scientific revolution. George Sarton was fond of saying that the history of mathematics should be the core of the history of science, just as the history of science should be the core of the history of civilization. However, mathematics is quite different from science; perhaps I need only call your attention to what is the nearest thing to a mathematical revolution that one can think of. Bertrand Russell liked to say, that the greatest discovery of the nineteenth century was the discovery of pure mathematics. If this was a revolution, then with it you have the introduction of non-commutative algebras and non-Euclidean geometries. You will notice, however, that whereas in 1600 you might have asked a scientist if he were an adherent of Copernicus or of Ptolemy or of Tycho Brahe, no one in 1900 would think of asking a mathematician if he were an adherent of elliptic geometry or parabolic geometry, or hyperbolic geometry. In other words, all three still are perfectly good mathematics, as are also both commutative and non-commutative algebras, whereas I think that we in this group will not claim that all three of the Renaissance astronomical systems are perfectly good science.

GERALD HOLTON: I would like to say very briefly what a pleasure it is to have heard these fine papers today, and to thank those who have arranged them.

A short remark about physics at Harvard in 1946 in view of the reference Professor Ravetz made to it. It was a good year. We were celebrating P. W. Bridgman's Nobel Prize for high-pressure physics, and younger people were hard at work on research that was presently recognized also. I just want to exempt the Jefferson and Lyman laboratories from having contributed to the situation to which Professor Ravetz addressed himself.

More seriously, I wonder whether Professor Settle's interesting paper does not raise a question. It has to do with the fact that he likes to hold on to such terms as "normal" and "abnormal", or "ordinary" and "extraordinary", and abandon revolutions and discontinuity. It probably can be done, although Professor Ravetz said, we may now be stuck with them. Yet, there may be an opportunity lost if one continues to use terminology which one believes to be basically wrong. For example, one might not be able even to ask certain questions that might be important.

THOMAS B. SETTLE: My own remarks were directed, not so much at reforming the language. However, I think Professor Holton is right. If my remarks turn out to have any validity, then Holton is right. The language eventually will have to be reformed.

STEPHEN BRUSH: I wanted to make some remarks partly based on my recent historical research, and partly on my own past experience as a scientist before becoming interested in history. At that time, I suppose, I was being a "normal scientist" in some sense. What I found myself doing a lot of the time was mathematical calculation. In listening to some of the discussions of how a revolution gets started, I wonder if perhaps those who write history have paid enough attention to the very technical apparently dreary aspect of what the scientist does in the world of mathematical calculations, when he is not worrying about "anomalies" that challenge his physical understanding.

I think it was Whewell who said that calculus, as compared to old fashioned geometry, is like going through a tunnel rather than going around the hill—you get more quickly to the other side, but you lose the view of the surrounding countryside. I think this is often what happens when a scientist is doing mathematical calculations, but I prefer the metaphor of diving under water. One has a certain problem to solve. One is presumably committed to a certain world view as we have been told about Einstein and about Copernicus, Planck, and others. One works very persistently on a particular problem. During this time one is presumably not arguing with oneself about the philosophical questions or even about the physics of the situation but one is simply solving this technical problem. You swim very hard without taking your head out of the water to look around, then, finally, you emerge on the other side. You have gotten a solution which is presumably conditioned by some criteria of simplicity, numerical harmony, etc., but you may find that the physical or metaphysical baggage that you started off with when you dived in has gotten somewhat waterlogged and perhaps needs to be discarded, although you do not know quite what to put in its place. You may have started a revolution without intending to do anything more than solve a technical problem defined within the old world view, yet the solution forces you (or at least your followers) to abandon that world view.

The examples with which I am familiar, primarily from early nineteenth- and twentieth-century science, some of the things that Tom Settle said about Galileo, and some of the things that have been said about Copernicus by Dr. Ravetz, suggest to me that there is often a period of persistent computational drudgery leading to the adoption of a new physical theory. The theory is *then* rationalized in a new framework, *after* it has been established mathematically and is found to fit some of the data.

Perhaps this is a way that one can understand how it happens that when looking at a period 25 years before, and 25 years after a certain event, one finds a fairly substantial discontinuity in the physical basis that is accepted, yet it is very hard to put a finger on a particular crisis or a particular time when a scientist has explicitly criticized or rejected a previous theory on *physical* grounds. Yet somehow it does get rejected and the change we are concerned with does happen. I would therefore say, that perhaps some of these mathematical calculations, which we tend to skip over since they are not very interesting to talk about, can play a very important role as a driving force in insidiously initiating revolutionary changes in science.

C. DORIS HELLMAN: Having heard a little bit about the use of mathematics, I think we might at this point make a correlation between the changes in scientific theory and the changes in the available mathematics. Kepler came awfully close to devising a calculus—I do not think we could handle the changing rate of motion necessary for the Newtonian Universe without the calculus. I do not think we could have had the modern cosmology without the non-Euclidean geometry.

WILLIAM R. SHEA: Professor Shapere made an important distinction between the cultural and the scientific revolutions in the Copernican innovation. He mentioned two criteria of scientific acceptability, unification and realistic premise, and he suggested that the realistic basis helped subsequent developments. Would Professor Shapere care to comment on this view—why a realistic epistemology was an aid to scientific progress, say over and above an idealistic or positivistic or conventionalistic epistemology?

DUDLEY SHAPERE: I do not think that is always true: a realistic interpretation of one's work is not a necessary condition for taking a revolutionary step in science. Often in the history of science, when a radically new idea is introduced, its author is so unable to accept its revolutionary consequences that he refers to his ideas as a mere "calculating device", not to be taken as a description of the way things are. Max Planck thought of his quantum hypothesis in this way, and there are parallel situations in the innovative work of Galileo, Newton, Maxwell and Schrödinger, to mention only a few examples. On the other hand, in the case of Copernicanism, a "realistic" interpretation did happen to ease the way in the manner I have suggested.

Thus, while in many cases a realistic interpretation serves the cause of innovation and progress in science, in other cases it plays no role. Perhaps, then, we should recognize such variety in the types of reasoning that occur in cases of major scientific change, and turn to careful examination of those types, instead of trying to construct a single unitary model of "scientific revolutions".

BENJAMIN NELSON: It would be a pity if a group such as this were to let its sessions end without making some provision for the future exploration of issues which need continuing study. In this connection it occurs to me to recall the example of Professor A. C. Crombie who has been sitting in our midst all too quietly today. In a remarkable volume which has hardly received the attention it deserves, Professor Crombie not only fathered a number of learned and challenging "position papers", he also took the initiative in encouraging the authors and Conference participants to identify themes and issues for future research. I trust that in the remaining day and a half of our current session all of us will give some thought to the agenda for our next gathering.

EDWARD ROSEN: Let me sneak this in before you go. This is a minute contribution to the debate between Professor Gingerich and Professor Nagel. Was there a scientific crisis or an astronomical crisis in the lifetime of Copernicus? Beyond his own statements about how he came to depart from tradition—statements which I think we should accept at face value—there are two other matters of fact to which I should like to direct your attention. These events did not make a social crisis, but they did make a personal crisis in the intellectual development of Nicolaus Copernicus. His teacher in astronomy at the University of Bologna was Domenico Maria da Novara, who in 1489 claimed that his determination of the latitudes of cities on the Mediterranean shore differed by about one degree from the latitudes contained in the *Geography* of Ptolemy. The conclusion was drawn by D. M. da Novara that the axis of the earth had shifted. It may well be that in Copernicus' imaginative mind a possible shift of the earth's axis could have led to the thought of a motion of the earth as well.

The second event to which I should like to direct your attention is this: in 1496, when Copernicus was a student in Italy, an *Epitomē* of Ptolemy's *Syntaxis* was published, and this was a book which he studied very closely. The title of that book is a little misleading because it was more than an *Epitomē* of Ptolemy's *Syntaxis*. For instance, at the end of the 22nd preposition of the 5th book, there is a little sentence which had, I think, a tremendous impact on the mind of the student, Copernicus. That sentence declares that if Ptolemy's theory of the Moon is right, then there must be times in the month when the apparent diameter of the Moon should be twice as large as at other times, and this is a phenomenon that nobody has ever observed. So here you have two fatal thrusts at the reliability of Ptolemy and it may well be that their combined effect led the young Copernicus to think that Ptolemy could be wrong in asserting the stability of the earth and in implying a doubling of the apparent diameter of the Moon. In that case he might well be wrong in other matters as well.

I wish to thank all the speakers for their stimulating papers, and the members of the committee who organized this Conference.

SESSION 4

The Humanistic Significance of the Copernican Heritage

Introductory Remarks

RAYMOND J. SEEGER

National Graduate University, Washington, D.C.

IT OCCURRED to me recently to ascertain from some books in my study what attitudes intellectual historians exhibit toward Nicolaus Copernicus. I found a wide range of views. In the first place, there has been an amazing amount of indifference from the time of Copernicus to the present. In many instances he is regarded as a typical renaissance man—indifferent to science. John Donne, the metaphysical poet and Dean of St. Paul's (London), grouped Copernicus with troublesome innovators in "Ignatius, his Conclave" (1611); but did not assign him preeminence in hell on that account. "What cares he (Lucifer) whether the earth travel or stand still? Do not men believe? Do they not live just as they did before?" More recently, Harvard's Perry Miller noted (1961) with respect to New England Puritans: "Not that they were more enlightened than their age about Copernicus and Galileo, they were more indifferent."

Strangely enough there are some who assign Copernicus a relatively minor role. Vienna's Friedrich Heer (1953) argues, "There is good reason for grouping Franck (Sebastian) with Copernicus and Paracelsus. . . . All three were deeply medieval figures." Columbia's John Herman Randall, Jr., insists (1940), "Following upon this discovery (by Copernicus) came the even more momentous Cartesian revolution. . . . Thus was the Copernican revolution consummated by Galileo; but even more significant was the Cartesian revolution which created a new physics" (questionable).

Finally, there are those who recognize the far-reaching significance of Copernicus' work. U.C.L.A.'s Will Durant, having learned with Ariel Durant from "Lessons of History" (1968) that "the initiative individual—the 'great man', the 'hero', the 'genius'—regains a formative force in history", concludes (1957), "If we judge greatness by influence—which is the least subjective test we can see—we may rank Luther with Copernicus, Voltaire, and Darwin as the most powerful personalities in the modern world." The former Secretary of the Smithsonian Institution, Leonard Carmichael, went further, "No other man has so widely influenced the thought of the world as did Copernicus." The nineteenth-century mathematical astronomer of the U.S. Naval Observatory, Simon Newcomb, observed, "It has always seemed to me that the real significance of the heliocentric system lies in the greatness of this conception rather than the fact of the discovery itself. There is no figure in astronomical history which may more appropriately claim the admiration of mankind than that of Copernicus." I myself, impressed with man's amazing technological achievement in bringing moon rocks to the Earth, am even more amazed at what he has learned about celestial bodies without the necessity of space travel.

Copernicus belonged to the two cultures, he was both a scientist and a humanist (in a broader sense than Charles Percy Snow's restricted intellectual phases of life). He has often been called the "father of modern astronomy", which is an ambiguous phrase in view of the nebulous meaning of the term "modern". Certainly Copernicus was a mathematical astronomer in the ever-developing tradition of Eudoxus and Ptolemy. He bridged antiquity and modern times with his use of planetary circles of early astronomy and the Sun-centered system of later astronomy. Galileo, indeed, shows in the frontispiece of his *Dialogue* (1632), Aristotle pointing beyond to Copernicus. Copernicus was particularly interested in the mathematics requisite for describing planetary motions, i.e. trigonometry, including spherical trigonometry. He was consulted by the Lateran Council about the proposed calendar reform. For 31 years he devoted himself privately and freely to seeking a physically real solar system. He was, however, prudent and certainly not aggressively self-seeking, as is evident in the Preface of his *De revolutionibus orbium coelestium* (1543): "When, therefore, I had pondered these matters, the scorn which was to be feared on account of the novelty and absurdity of the opinion impelled me for that reason to put aside entirely the book already drawn up."

Humanism is often used by self-designated humanists in a narrow sense. I prefer a broad one such as that of the Roman playwright Terence, who claimed (163), "Homo sum; humani nil a me alienum puto" (I am a man; I count nothing human indifferent to me). Or, in the words (1873) of the English critic Walter Pater, "The essence of humanism . . . nothing which has ever interested living men and women can wholly lose its vitality." In this spirit, I believe, Copernicus was truly a humanist. Having studied law at Bologna and Ferrara, he was awarded (1503) a doctorate in canon law; he had been made (1497) a canon of the cathedral in Frauenburg, Poland. Here, he served primarily as a church administrator. In this capacity he was responsible for property, including the defense against attacks by the Teutonic Knights (1519–21) and currency reforms at the request (1526) of Sigismund I of

Poland (he was the first to formulate the monetary law enunciated later (1550) by Sir Thomas Gresham). He was concerned also about people; he served frequently as a physician, having studied medicine at Padua (1501–3). Thus, from the viewpoint both of the character developed in ministering to public needs and of the thought nurtured in privately seeking to understand celestial phenomena, he was truly a humanist.

With respect to Copernicus' influence, it is probably more appropriate to speak of a cultural revolution rather than a strictly scientific revolution. Then as now, to be sure, there was considerable social inertia. In connection with the 400th commemoration of Copernicus' death the Nobel physicist James Franck remarked, "The contemporaries of Copernicus did not like to give up the idea that they lived in the center of the universe since it flattered their ego, just as nowadays people indulge in wishful thinking and are prepared to accept the most absurd theories if they are apt to inflate their feelings of importance." He cited astrology as a current example. How much more true today when the so-called space age is commonly called the age of Aquarius!

The most significant aspect of this cultural revolution was undoubtedly the philosophical and theological implications of Copernicus' theory, leading to a truly intellectual and spiritual revolution. For example, "A fearful shock to the accepted Christian faith was given by the promulgation of the Copernican hypothesis". William Inge, the gloomy Dean of St. Paul's, pondered that "the church of the 20th Century has not yet digested the Copernican hypothesis". Philip Melanchthon, the Lutheran scholar, ridiculed Copernicus as "that Sarmatian astronomer who is trying to stop the sun and move the earth". His words boomeranged as a truthful description; today they are inscribed on the high altar of the church of the University of Cracow where Copernicus first learned humanistic astronomy from Albert Brudzewski. To be sure, like all infections developing about impurities, false thinking had to be removed so that natural healing might ensure, but scars have remained to this day. Durant, admitting that "Copernicus . . . did most damage to medieval Christianity", emphasizes that "the heliocentric astronomy compelled men to conceive God in less provincial, less anthropomorphic terms; it gave theology the strongest challenge in the history of religion. The Copernican revolution was far profounder than the Reformation." Carmichael observed hopefully, "He (Copernicus) became one of the beneficient human liberators who opened up our modern world of truth and freedom. The Copernican revolution is not ended." Walter S. Adams, late director of the Mt. Wilson Observatory, concluded, "Copernicus joins also that selfless company whose unending quest of nature reveals the power of the human spirit to transcend earthly bond and its own brief hour."

"The Humanistic Significance of Our Copernican Heritage" is the title of the address for this evening (December 27, 1972). In the preface of his *Astronomy and Cosmology in the Achievement of Nicolaus Copernicus* (1965) our speaker confesses:

Raymond J. Seeger

> I teach the history of science at the University of Leeds and in my introductory course a central place is taken by the Copernican revolution. It seemed to me that such a phrase would be nearly meaningless unless one had some idea of what Copernicus achieved and why he devoted his life establishing astronomy on the basis of the motions of the Earth. When I sought the answers to these questions in the many excellent books written in the tradition of the history of scientific ideas I found nothing which claimed to be even a plausible reconstruction.

A native of Philadelphia, a graduate of mathematics and physics from Swarthmore College, a Fulbright scholar with a doctorate in pure mathematics from Cambridge University, now senior lecturer at the University of Leeds, our speaker is Jerome R. Ravetz.

The Humanistic Significance of Our Copernican Heritage

JEROME R. RAVETZ

Leeds University, Leeds, England

IT IS a great pleasure and honour for me to be here this evening and to be brought over from England as a guest speaker by the American Association for the Advancement of Science. But I hope I do not give offence to my hosts if I say that the occasion which has brought us together, the commemoration of a great scientist and a great man, is for me the source of an even greater honour for my small contribution. For with the fifth centenary of the birth of Nicolaus Copernicus, we review our indebtedness to that one man who, alone and single handed, could successfully bring science and the scientific method to bear on an outstanding philosophical problem, perhaps the greatest single problem of man's position in the world of Nature. Copernicus was one of the great intellectual and moral giants of our civilization; his achievement in effectively proposing the motions of the earth had so many consequences in human thought and practice, that his heritage to ourselves (our topic for tonight) could be summed up as—ourselves.

To put the point that way, and to take it seriously and not as a mere rhetorical flourish, might seem to stop the conversation. Were I a perverse and ingenious student writing an examination essay on this topic, I might be tempted to make such a one-word answer, and hope that my wit would be appreciated more than my impudence resented. But the risk would be high, especially since the examiner could write in the plentiful blank space: "The answer is banal; the candidate fails to see that the question actually calls for an explanation of 'ourselves' as heirs of Copernicus."

For what we are is, to a great extent, what Copernicus and his successors created. Hence, as historians, we study the past partly for the same reasons as anybody seriously studies anything: to know ourselves. The statement "Copernicus is the first great modern scientist" may seem to be about the past. But it has a converse form: "Our own ideas of greatness in modern science are largely formed by the achievement of Copernicus". Hence when we look at *our* Copernican heritage, we are attempting to learn something about that topic of perennial and absorbing interest: ourselves.

Copernicus the Man

We and our precious concerns can wait for a moment; let us first look at this man in whose memory we are assembled tonight. But for one single memorial, his career might well have been indistinguishable from that of dozens of others that started in 1473. We have the family of good standing, the early death of the father, and the adoption by an uncle; family and education are divided between the town of Toruń (Figs. 54 and 55; p. 153) and the capital of Kraków (Figs. 56 and 57; p. 154). Although situated at the far Eastern extremity of the Europe of learning and letters, Kraków was no frontier town. Its surviving churches and palaces attest to its Royal magnificence, and the University had close connections with humanist circles in Italy.

Given a start on a public servant's career by a sinecure and a canonry, our young man headed for Italy in 1496 to a distinguished University, Bologna (Fig. 58) for professional training in canon law. A second Italian tour (Padua, 1501) was for medicine, and also for the "doctorate" which seems to have been as much a union card then as it is now (Fig. 59). But student life cannot last for ever; by 1503 our young man returned home, to join the personal staff of his uncle, now the Bishop of Varmia on the far north-east Baltic coast of Poland near Gdańsk (or Danzig). After some time he left this post, perhaps foregoing preferment thereby in order to participate actively as one of a collective leadership of canons of the cathedral. Their concerns were not the peaceful tasks of arranging services and keeping the choir boys in order. No, all the talents of an administrator and statesman were required, for Varmia was an exposed land in troubled times: the Teutonic Knights were a decaying but still very menacing order, and Luther's Reformation had its repercussions there as well. There then passed some thirty busy years, at whose end he was an old man; and he died.

On his death-bed he handled a book, brought from Germany, that was his guarantee of the immortality of his name. For this man of affairs had long been known to have a hobby, astronomy; and indeed to have what the English call a hobby-horse, a crazy theory that the Earth spins around once a day and also goes round the Sun once a year. Most tantalizing, he had a large book in which he simultaneously proved these motions of the Earth and also brought a much-needed reform to the science of astronomy. His astronomical prowess had been recognized from his student days; he had assisted a distinguished professor at Bologna and had lectured at Rome in 1500. Moreover, there was in circulation a brief outline of his world system; he corresponded with his colleagues on astronomical questions; and he had built observatories and instruments at various places. But one thing he would not do was to publish his great book; he was willing to publish astronomical tables derived from his theories, but the theories themselves were not to be revealed for public incomprehension and ridicule.

So his work might well have died with him, had it not been for the fortunate intervention of a young German Lutheran, Rheticus. Establishing himself as Copernicus' disciple and public-relations man, he secured a printer's copy of the great work, and got it started on publication. This was the book that was brought to the dying Copernicus. In this book, soon distributed all over Europe and twice reprinted as a scientific text, were contained the seeds of the cosmology and the science that have made our world what it is.

That this man of one book was an intellectual giant, there is no doubt. His moral qualities may not have appeared so clearly from my brief sketch, or indeed from some other more extended accounts. But remember: for most of his life he was, to his knowledge, the only living soul who knew that the Earth is not at the centre of the cosmos. "All the world is wrong, and I am right": this conviction requires courage. It requires stamina too, when its possessor is not sheltered in a scholarly community where all sorts of eccentricities are tolerated, but instead is fully engaged, in companionship with others, on very immediate and pressing mundane tasks. And one further feature of Copernicus' situation makes him truly a hero. In the face of the massive arguments *against* the Earth's motion from accepted natural science, from sacred Scripture, from obvious common sense, what could he counterpose? Heuristic principles, and apologies for a physics he could not produce. *Where* was that conclusive argument for the Earth's motion, that would override all objections and quiet all doubts? It is not to be found in the writings of Copernicus, nor indeed in those of his immediate successors.

Considering Copernicus the man, committed in full to an idea fraught with such insuperable difficulties and hazards, we naturally ask how he came to be in such a situation. What was the evidence that was conclusive for him, that made him decide to devote his great talents to proving his hypothesis that was certain to appear odd and indeed abhorrent? We can only guess. My personal guess has involved an astronomical problem of a depth commensurate with that of his cosmological revolution. Endeavouring to avoid viewing Copernicus backwards through Galileo's telescope, I have tried to reconstruct a natural sequence of problems that would bring him along from work on the foundations of astronomy, through to the fundamentals of cosmology. These problems were those of the reference-frames, of observable position and of time, whose definitions must be internally coherent and operationally feasible, if there is to be a genuine science of quantitative astronomical observation. My reconstruction of a possible first phase of Copernicus' path has rested on a distinction between "the motions (in consequence) of the stellar sphere" and "the precession of the equinoxes". The former was the traditional way of explaining two related phenomena: the slight difference between the equinoctial and the sidereal year; and the slow changes in longitude of the fixed stars. From this earliest writing, Copernicus chose the latter concept; and the two modes of description are definitely not equivalent. From this non-equivalence, Copernicus could have been driven to suppose the rotation of the Earth, as the only rational way to save the most important phenomena then accepted as irregular.

The details of this hypothetical reconstruction are still a matter for further study and debate; and they are not our concern tonight. But if I may be permitted just one remark on more recent evidence on this matter, I would like to mention the records of the Kraków school of astronomy in the fifteenth century. This was as strong, and as famous, as any in Europe. Its teachers wrote excellent monographs on mathematical astronomy, and its students went on to distinguished careers in astrology. Polish scholars are now sorting and analysing these sources, as part of their great international co-operative endeavour for the Copernican year. I am told that they are finding a continuity of techniques and also of problems, between the Krakówian masters and their most illustrious pupil. Hence the hypothesis that Copernicus was led to his cosmological revolution in an entirely natural way while still a student at Kraków receives, if not corroboration, then at least enhanced plausibility.

Regardless of the details of Copernicus' path to his conviction of the motions of the Earth, we know that he got it early in his career and held to it steadfastly. He devoted himself to its establishment, through decades of work and distraction, into the declining years of his life. He continued this endeavour in the absence of nearly all the supports that ordinary men need to sustain a lively faith. He could not even be an eccentric. This man had what it takes: grit. Thus this Polish Cathedral canon, a humanist and scholar in his spare time, kept the faith with his very personal conviction of the structure of the Universe and passed to his successors the materials for both the reform of astronomy and the revolution of cosmology. The direct consequences of these both extend to our own time, and constitute our Copernican heritage: let us now consider them.

The Copernican Heritage

The point that is easiest of entry, best known, and still most fundamental for science, is that of astronomy. Before Copernicus, there was a science of the heavens, which was integrated with a natural philosophy of space and of place, and also with a general view of the whole world as a cosmos, alive and dramatic. Copernicus did not directly challenge this whole scheme of reality, which was no less self-consistent, in the context of a body of accepted facts and methods, than is our own. He worked on a more restricted front, where technical competence rather than metaphysical illumination would prove decisive. The Copernican heritage was kept alive through the crucial early decades after 1543 by the achievements of the *De revolutionibus* as an astronomical treatise. Later astronomers who found Book I embarrassing or even threatening, were unable to dismiss it from view. For it was physically bound together with the technical chapters, in which were laid the foundations for a new reformed astronomy, and which were the indispensible source-book for any intending astronomer over the next 60 years. For in this first phase of the Copernican revolution, there was no entrenched old guard vigorously defending their paradigm. I am sorry to have to tell you this, but the common image of "Ptolemaic" astronomers adding an increasing burden of epicycles to their planetary models, and so creating an urgent need for simplification, is simply wrong. I have never seen any reference to an astronomical treatise of the fifteenth or early sixteenth century in which such hyper-elaborated planetary models are used. I request anyone who believes in the multiplicity of epicycles in pre-Copernican astronomy, to supply me with such a reference. It is far more likely that Copernicus considered the entire structure of mathematical astronomy to be in a confused and barely competent state. This was certainly the impression of his successors, concerning the situation he remedied. That is why they called him "the restorer of astronomy", "the Ptolemy of our age", even when they averted their gaze from his cosmology.

Also, it was on the technical front that the later heirs of Copernicus scored their victory. Out of the irrefutable facts of astronomical observation, and unchallengeable conclusions of mathematical deduction, there eventually resulted a veritable revolution in the whole system of the world. The Earth came to be seen as a massive body in mutual accelerative attractions with all the others; and in these terms it was possible to take mathematical astronomy to a previously unimaginable degree of power and accuracy. This overwhelming success of an astronomy derived from Copernicus was the final, conclusive argument (in the eighteenth century) against those cautious and pious individuals who tried to the last to treat his cosmology as mere hypothesis.

The technical astronomy of Copernicus was a matter for specialists; the wider interest and debate focused on its cosmological consequences. In the writings of Copernicus himself these were kept as modest as possible. But gradually the new cosmology developed itself, culminating in the complete transformation of our conceptions of space and place. The closed, symmetrical earth-centred, indeed man-centred cosmos of all earlier traditions, eventually gave way to an empty Universe of vast extent, with our own Earth as a minor planet of a minor star of a minor galaxy.

This deep change got underway in the seventeenth century; its prophets were Descartes and Galileo. We should not assume that their vision was a necessary and simple consequence of Copernicus' work. As Pascal said, it was the *emptiness* of those vast spaces that frightened him; and we should recall that Bruno and Kepler both had Copernican universes formed or filled by some sort of intelligence or activity. But, up to now at least, we have looked on Descartes and Galileo as the true heirs of Copernicus, and interpreted their conceptions as further moves in a direction established by him. Whether Copernicus himself would have accepted the cosmic gloom of Bertrand Russell's classic statement in *A Free Man's Worship*, we do not know; nor indeed can we be sure, that it is the final word on the subject. But the naïve anthropomorphism of the earlier European system of the world is now gone beyond recall; this is the essential core of the cosmological achievement of Copernicus.

The displacement of man as measure of the natural would soon extend beyond cosmology, into natural science in general. The accepted principles of scientific explanation had previously all been anthropomorphic in one way or another: taking basic categories from those of sense-experience, and taking causal agencies from those of purposive actions. Thus the pre-Copernican world was conceived to consist of elements derived from ordinary perception: earth, water, breath and fire. These were related by the fundamental paired perceptual qualities; hot–cold, wet–dry. Other, derivative qualities were also in polar opposites, and their persistence in common speech after all these centuries of scientific refutation is a testimony to their strength; for example, heavy–light, fast–slow, light–dark. The principles of change were equally close to human experience. The alchemists spoke of the "perfection" of a base metal, while Aristotle had "natural" motions of bodies to their "natural" place, as contrasted to "violent" motions. In the explanation of any complex problem, use could be made of such maxims as "nature does nothing in vain". This whole approach to science is not so infantile as it can be made to appear—and was, in the seventeenth century. For it was originally used to give a qualitative plausible explanation of ordinary phenomena as perceived through the senses. Only after a centuries-long detour through the world of abstractions, either of controlled experience in the laboratory or of theoretical constructions, can science now address itself to those same questions, such as, what makes thunder and lightning?

Although the physical principles cited by Copernicus himself do no more than hint at the "dehumanization" of the realm of scientific explanation, still this process can be seen as taking its origins in his work more than in any other of its time. And to the extent that modern natural science has developed up to now by successive advances in "reduction" or as some would call it "objectivity", then it is the product of the work of Copernicus.

There is yet another aspect of our Copernican heritage, whose humanistic significance is perhaps the most basic of all. For what we find, out in the world, depends to a considerable extent on what we expect to find. This is not to say that we create what is out there; but rather that there is a process of selection and elimination. It is very difficult indeed to find what one is not at all expecting to find, especially if the "finding" involves long programmes of research organized around the discovery of a particular class of effects to the exclusion of others. In practice this has meant that since the time of Copernicus men have been progressively less concerned to find wonders and prodigies in natural phenomena, less interested in rumours of miracles and in old wives' tales, and increasingly concentrated on the elucidation of the impersonal, regular workings of Nature. By his silent example, Copernicus was both a pioneer and a leader in this progress. For unlike most of his predecessors, contemporaries and immediate successors, in astronomy and medicine, Copernicus did not practice astrology. If he did have a metaphysics serving to guide his early speculations towards a heliocentric universe, it seems likely to have been compounded out of two elements, a humanists' neo-Platonist Sun-worship, together with a mathematician's commitment to the rationality of the system of the created world.

Hence in that progress of the human spirit away from what used to be called "Medieval", that is rule of men's minds by dogma and superstition, Copernicus was an illustrious humanistic pioneer. He ranks with such contemporaries as Erasmus and the young Thomas More. And since the final victory in that particular struggle is only now passing out of living memory in even the most enlightened centres, this aspect of our Copernican heritage, the most directly "humanistic" of them all, is to be kept uppermost in our minds on this anniversary.

Thus we may sum up the Copernican heritage as scientific, philosophical and humanistic. A great scientist laid the foundations of a revolution in philosophy, and thereby contributed in many ways to the cultural development of humanity. Five hundred years after his birth, this man of whose life and personal endeavour we know so very little, speaks to us by the example of his scientific brilliance and philosophical courage.

Our Copernican Heritage

Having been reminded of the man and his achievement, we can now ask, just what does he say to *us*? What indeed is the humanistic significance of *our* Copernican heritage? I do not raise this question out of a belief that anyone here this evening, or indeed any scientist of our generation, can have an achievement, or even experience a challenge comparable to those of Copernicus. But the challenges of our time, and the achievements to which they may give rise, are the ones we have. Dwelling on the glories that date from 1473 may give us only a superficial and false reassurance and guidance for the tasks that lie ahead in 1973. And so I would like to have another look at this Copernican heritage, to see more precisely how it speaks to us.

The first aspect that strikes me is a simple quantitative one: length. Not so long ago we celebrated the three-hundredth anniversary of the foundation of the Royal Society of London. Then we celebrated the four-hundredth anniversary of the birth of Galileo. With this year, 1973, we reach the first quincentenary. If we date "modern science" from this point, or perhaps wait two or three decades for a commemoration of the composition of the *Commentariolus* (the first statement of the heliocentric system), we have the first half-millenial bench-mark for "modern science". That is a long time. Perhaps we should stop using the term "modern" for something that stretches back so far. Let's call it "European", with no disrespect to my hosts; we are all part of a certain phase of European civilization, whose period of continuous development, stretching back from the present can be considered as just a bit more than a half-millenium long.

A lot can happen in 500 years. In ancient times great empires could go through full cycle from obscurity to disaster in far less time, and few survived even that long. Other great movements in civilized culture, in letters, art, or religion, have never kept their vitality unimpaired for such a long period. Yet Copernicus speaks to us through a continuous succession of zealous disciples and great masters. Although every nation has seen the scientific spirit quicken and ebb alternately, European science as a whole has progressed steadily from his time to ours. Through such a long period it has kept the identity of its essential nature: that is, after all, why we are here in commemoration of Copernicus. But the external circumstances of sciences have, as we all know, changed enormously. On this occasion there is no need for me to go into great details in describing the new responsibilities that have been thrust upon the community of science in these past few years. But it is appropriate for me to review the ways in which this new situation has made *our* Copernican heritage different from that of earlier times.

Up to quite recently, the achievement of Copernicus in establishing a tradition seemed quite clear and simple. Stemming from his work, there came progress on many fronts: in science, a progression from error to truth; and in human affairs, from mysticism to logic. Yet this optimistic view of the process was derived from a particular and temporary view of the future, of the present and of the past. When I and my colleagues look again at the records that indicate what it was like while it was still going on, the picture that emerges is more complex and less comforting.

The study of our Copernican heritage has led me to the following uncomfortable thesis. We can no longer maintain the easy identification of the application of correct methods with the attainment of true facts. The great achievements of science are not won solely by the use of the unbiased eye and the unprejudiced mind. And the recognition of the true amidst the false is, in science as in life, progressively more difficult in proportion to the importance of the matter at hand. Thus the foundations of the simple faith in science of earlier years will no longer do.

I hinted at one of these difficulties when I mentioned earlier that we are not sure of the grounds on which Copernicus became convinced of the truth of the heliocentric system. For, as we look back at it now, there was not available to him any correct observational data which made the new cosmology either certain or even overwhelmingly probable. Then we may ask, when did such data come in? While it is true that discoveries by Galileo and others rendered the popularized Aristotelian cosmology progressively less plausible, no single one of them, or group of them taken together, provided a "necessary demonstration" whose conclusion was a moving earth. Galileo is likely to have been aware of this problem; hence his great commitment to his theory of the tides, based on visible, mechanical motions.

It is true that in this scientific debate, the parties all recognized what would count in principle as conclusive evidence. One sort was a stellar parallax, and this is handy because its presence would be conclusive for Copernicus, while its absence could be argued away. This was just the tactic of Digges; he searched for parallax and instead "found" an infinitely extended universe. Stellar parallax was eventually found, and genuinely so; but this was already in the nineteenth century, by which time it did not matter any more.

This example raises two questions. First, when, and on what grounds, was the Copernican theory completely accepted as scientific fact? Second, what principles of scientific method *should have* governed the reception and evaluation of the new doctrines in the intervening period? On the first, it seems that Newton's *Principia* did the job. As a rational theory of the motions of the heavenly bodies, it overwhelmed all opposition; and its arguments were quite strictly inconceivable within a geocentric system.

Before then, there was a welter of evidence, whose weight steadily inclined, through the seventeenth century, to the Copernican side. If we consider the methodological question during this period of flux, we find ourselves with some hard choices. On the one hand we can say that the Copernican theory was only an hypothesis with more or less probability, and hence Osiander and Bellarmine were right, and Copernicus and Galileo were wrong. If we dislike this judgement, as a betrayal of great scientists and their vision, we seem to be forced close to an embarrassing alternative: that scientists should receive credit for the subsequently confirmed correctness of their beliefs, regardless of the scientific justification for their own adherence to them.

This sort of judgement is not an idle game but a basic problem in the interpretation of past science. In this connection, we can construct some paradoxes that are no less real than troubling. What would a reputable publisher today do with a treatise on a controversial problem, if the referee reports that the basic positive argument is incomprehensible, and that the discussion systematically avoids mentioning the theory most favoured by experts in the field? He would not be likely to hail it as a classic of science. Yet this is what we now do with Galileo's *Two World Systems* in spite of its flaws in just these respects: the theory of the tides is very difficult, to say the least; and Tycho's system is ignored.

We might give ourselves some comfort in the face of such paradoxes by recalling that during scientific revolutions the normal rules of science cannot apply strictly; and hence such paradoxes creep in. But we now know that advances in sciences are humanly interesting and philosophically significant just to the extent that they are revolutionary. Hence the rules of "scientific method" and the cautious "scientific attitude" would now seem to function more as constraints on mediocrity than as channels for creativity. To subsume this paradox under some general rule of human behaviour, we might need to quote William Blake's aphorism, "One law for the ass and the lion is tyranny." Or, in a particular example, we may say that we *know* Galileo was a genius and not a crank, precisely because he *broke* the formal rules of scholarly inquiry.

Once we lose that old certainty in science as the straight and wide Royal road to knowledge, we should expect even more uncertainties on the philosophical aspects of the heritage of Copernicus. At this point I need indicate only one of them, by the example of Tycho Brahe, the great Danish astronomer. He is remembered as an observer of unparalleled diligence and dedication, and also as a respectful critic of Copernicanism, who developed his own compromise cosmology. So far he would seem to be the perfect cautious scientist, wanting to get the facts in before committing himself to daring and implausible hypotheses. But we learn from his autobiography that the motivation behind all his observing was—astrology. Here he was not cautious; his faith in astrology and also alchemy was complete. It was the basis of his career, for an early successful prediction had established his fame and brought him patronage. But he was keenly aware of the insufficiencies of the quantitative data-base of this most important social science; and for this reason he engaged on his heroic labours of observation and computation.

Thus, this one man can be interpreted as a strict positivist scientist, or as a superstitious stargazer, depending on which part of him you choose to study. If we try to rescue him by saying that in the late sixteenth century it was still "rational" to practise astrology, we must then ask when it ceased to be rational, and on what grounds. I think you will find that just as the Copernican hypothesis came gradually to be accepted as fact because of its coherence with indubitable scientific theories, so astrology came gradually to be rejected as a pseudo-science, because of its incoherence with a new common sense of the world of Nature and man's place in it; it was never refuted, and could not be, any more than could, say, micro-economics.

Copernicus and Humanity

By this point in my discourse, I may appear to be deliberately and wilfully introducing a note of uncertainty and gloom into an event that should be one of pleasant commemoration of one of the greatest triumphs of science. I would deny this accusation, on both counts. First, I am not being wilful, I am only telling the conclusions of historical investigations as they have presented themselves to me and to many colleagues. You see, scholarly history is really as subversive a subject as philosophy: the latter exposes any conveniently inconsistent thinking about ourselves and the world, while genuine history corrects our tendencies to reconstruct the past in a self-congratulatory image of the present. And at the end of the fifth century of its continuous progress, European science should be able to survive (and perhaps even benefit from) critical scrutiny of its historical foundations.

For a long time our knowledge of Nature seemed, or should I say needed to seem, superior to all other sorts of knowledge held by men; perhaps even superior to those men themselves. But that was when the Copernican heritage was still young and raw, when it was still engaged in a hard struggle against dogma and superstition. But, in our lands at least, the Copernican revolution is an old one; let us not proclaim our filial loyalties to that revolution for so long as to become reactionaries. In this quincentenary year, we cannot plausibly claim to the omniscience or omnipotence, or indeed to the perfect beneficence of science. So let us not try to do so. If we accept some legitimate humiliation, then our reward is a lightening of the burdens of responsibility. We can be gratified that our scientific tradition is a glory of civilization and one of its main defenders; but we can now abandon the proud boast that science *is* civilization. We, and the world, are better off without such claims.

I would also deny that I am gloomy. When we see how much great tragedy, high drama, human error and plain muddle is involved in scientific progress, we are bringing it back to a human scale. So long as the great scientists of the past were predestined to anticipate the verities of the present, the whole process was as dehumanized as any textbook, and twice as frightening. Now we can and should admit that scientists are, and always have been, human. They make mistakes, singly and collectively.

They cannot fabricate a magic bullet that will protect mankind from his enemies, nor can they produce a machine that will enslave the human spirit. They, and their achievements, are part of the human comedy; they have laughter and tears.

This is for me one essential aspect of the humanistic significance of our Copernican heritage. The Copernican revolution itself was born in secrecy sometime in Copernicus' youth; it was nurtured in obscurity in a manuscript in a Baltic Cathedral chapter house; it was launched under a misapprehension by the Osiander preface; it was subjected to literally fantastic interpretations by such notorious disciples as Giordano Bruno; it was made the occasion of the most classic blunder in the intellectual relations of the Roman Catholic Church; and its thesis eventually became a commonplace fact, questioned only by barbarians and small children, in spite of running contrary to all direct sense experience. Yes, it was a triumph; but a very human one. The knowledge is of the impersonal world of Nature, but the knowers are men.

For the other essential aspect, let me remind us why Copernicus was great. Not because he happened to be the first man in modern times to know that the Earth goes round the Sun. Rather, because he earned the right to know this, using all his human talents in the face of every impediment and discouragement. We may think of him as devoting his life and talents to a single-handed attempt to climb an unscaled mountain, and one that turned out to be infinitely high. Recalling his endeavour as well as its success, we can be humble in the presence of his example. We can also be grateful that it has been provided for our own benefit.

FIG. 54. Map of the "Copernicus Country" in 1466.

FIG. 55. Toruń. (Courtesy of the Academy of Sciences, Warsaw.)

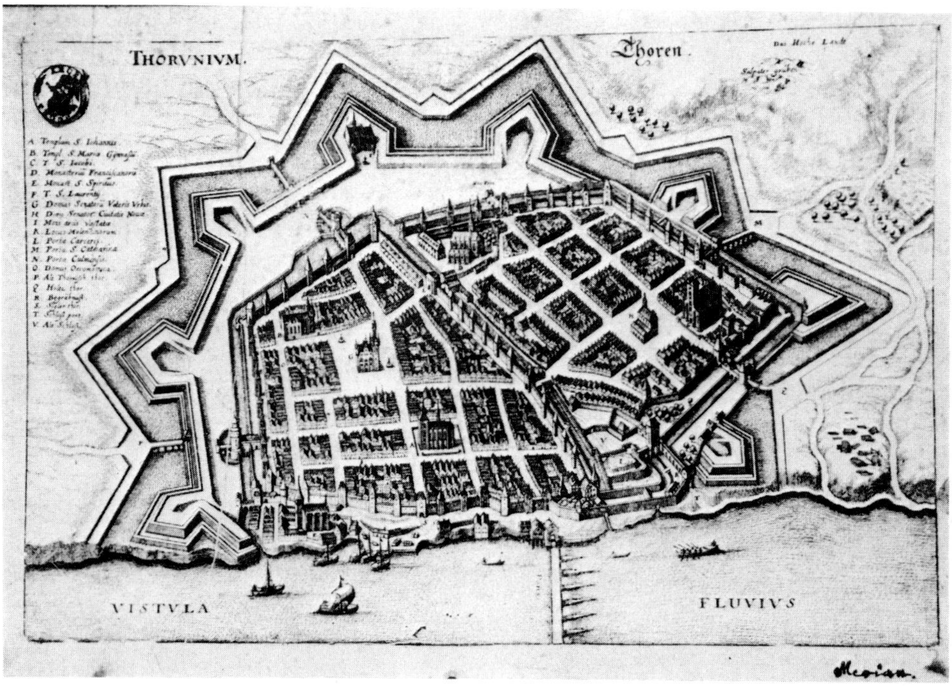

FIG. 56. Krakow. From Hartmann Schedel's *Nuremberg Chronicle* (1493). (Courtesy of the British Museum, London.)

FIG. 57. The central courtyard of the Collegium Maius where Copernicus studied as a college student in Krakow. (Photographed by Owen Gingerich in August 1972.)

Fig. 58. Bologna. From Hartmann Schedel's *Nuremberg Chronicle* (1493). (Courtesy of the British Museum, London.)

Tercia etas mundi

Padua

Patauina vrbs ytalice vetustissima celeb
rrimaq;. quã anthenorez Troya pfugũ
cõdidisse virgili⁹ testat; hic tamẽ ille vr
taq; fuerũt menia. qbus triplici circuitu ⁊ muro
patauiũ cõmunit. Et licet p eam Timau⁹ fuerit
delapsus. mltis tamẽ varijsq; fossis magno ⁊

FIG. 59. Padua. From Hartmann Schedel's *Nuremberg Chronicle* (1493). (Courtesy of the British Museum, London.)

SESSION 5

The Impact of Copernicus and His Work

Introductory Remarks

JOVES SOLDACZUK

Economic Councillor, Polish Embassy, Washington, D.C.

IT IS a great honor and privilege for me to present before this distinguished audience, on behalf of the Polish Ambassador to the United States, the introductory remarks at this session of the Copernicus Commemorative Conference.

May I also take this opportunity to express our highest regard and gratitude to you, Mr. Chairman, to members of the organizing committee and to all those who gave freely of their time and effort to prepare and organize this Conference.

I would also like to express our gratitude to the distinguished scientists from the U.S.A., and other countries, for their attendance and participation in the Conference, which contributes significantly to the world-wide observance of the 500th anniversary of Nicolaus Copernicus—one of the greatest Polish astronomers and scientists, whose impact on modern astronomy and physics is well recognized.

Since I am not an astronomer or physicist, I do not intend to discuss the scientific achievements of Nicolaus Copernicus. It will be done here by the most prominent specialists in these fields.

Let me concentrate briefly on some important biographical facts of the life of Nicolaus Copernicus and on the more general aspects of Copernicus' contributions to human sciences.

Nicolaus Copernicus was born in Toruń in Poland on February 19, 1473 in a burgher's family. Seventy years of his life, and 50 years of intensive studies and practical activities in many fields, were concentrated in a period of great transformation and transition between several epochs—when the feudal system was breaking up, when the humanist rejection of medieval ideals spurred interest in ancient culture, when reformation in religious beliefs was under way, and when human problems began to take place.

The Copernican Theory was developed when Poland was experiencing her "Golden Age" in economical, political and cultural life at the end of the fifteenth and during the sixteenth centuries. It thrived in an atmosphere of the progressive aspirations of the Polish citizens.

Toruń, the birthplace of Copernicus, was at that time one of the richest towns in Europe, profiting mainly from overseas trade with Western Europe.

Nicolaus Copernicus' father was a merchant involved in trade and politics. He died when his son was about 10 years old. Since Copernicus' maternal uncle took over his support, he was able to enter the Academy of Krakow in 1591. The Academy of Krakow was more than 100 years old at that time and was famous not only for its theology but also for its natural sciences, mathematics, and astronomy, with Wojciech Brudzewski as the most prominent among the astronomers. The Renaissance ideals were established early at the Academy of Krakow, long before Nicolaus Copernicus began his studies there. It was at the Academy that he first became imbued with the ideals of the Renaissance, and it was there also that he became interested in astronomy.

Studies at Italian universities during the period of 7 years (1496–1503) drew him deep into the current of humanistic aspects of the Renaissance. In Italy he studied canon law at the Universities of Bologna and Ferrara, and medicine at the University of Padua. Above all, he continued his studies and observations in astronomy regardless of his other interests, and lectured on astronomy in Rome in the year 1500 while studying canon law.

Copernicus returned to Poland in 1503. As a canon, he first worked for his uncle, the Bishop of Warmia (Ermland), as a secretary and physician. He was also involved in political and administrative problems of the Warmia province. As a canon of the Warmia Cathedral chapter he was chancellor, inspector general and administrator of the chapter's estate. During the war against the Teutonic Order of Knights he distinguished himself in the defense of Olsztyn—the capital of Warmia—and demonstrated his diplomatic talents and skill. After the war he was involved in economic reconstruction of the province and proposed a currency reform. In connection with this he wrote a treatise on ways of minting coins, in which he formulated a law which may be briefly summarized as: "Bad money drives out good." After 30 years that law was rediscovered by the English economist, Sir Thomas Gresham, and is now known in economic literature as "Gresham's Law". In Poland it is called the "Copernicus–Gresham Law" and Copernicus is recognized as the founder of Polish Economics (Fig. 60 on p. 162).

All of the numerous interests and activities of Nicolaus Copernicus show that he was not only a great Renaissance thinker but also a "man of action". Despite his intensive public activities, he did not interrupt his astronomical studies and observations.

In 1507 he finished a small manuscript entitled "Commentariolus", which was a brief exposition of his heliocentric theory. In 1519 he finished writing Book One, which is the most important part of his fundamental work *De revolutionibus orbium coelestium*. The whole work was completed in 1530, during the latter years of his life. In this work he developed the heliocentric theory of the structure of the universe which opened up not only a new era in the history of astronomy, but new horizons in philosophy and moral science and, to a great extent, contributed to other modern sciences. Figures 61, 62 and 62A (pages 162–3) illustrate present-day appreciation of Copernicus.

Looking at the life, work, and achievements of Nicolaus Copernicus from a modern point of view, one can say that he is symbolic of scientific courage evidenced by his opposition to traditional authority and dogma. Professor Sobolewski, one of Poland's most prominent scholars in the field of cultural history, expressed his view on this point by stating "not only were the teachings of Copernicus an act of courage because he contradicted scholastic traditions, but they were also an act of great intellectual courage because they contradicted the sensory data of general human experience".

We can regard the work of Nicolaus Copernicus as a model of true scientific research, free from traditional experience, habits, limits and dogmas.

FIG. 60. Coins used in Poland about Copernicus' time. (Photographs by Charles Eames.)

FIG. 61. Krakow, 1972: Copernicus celebrations. Photograph by courtesy of Klaus J. Kallabis. (*Die Zeit*, Hamburg.)

FIG. 62. Catowicze: Copernicus Monument. Photograph by courtesy of Klaus J. Kallabis. (*Die Zeit*, Hamburg.)

FIG. 62A. The Copernicus statue in the "Bastionsgarten" in Salzburg—a sculpture due to Thorak (1889–1952) who lived in Salzburg in his last years. (Courtesy of Dr.Ing. Dr.h.c. A.-F. Staats, Hanover.)

On the Quest of Direct Proofs of the Earth's Motions

ALEXANDER A. MIKHAILOV

Pulkovo Observatory, Leningrad, U.S.S.R.

IT IS often said that Copernicus proved the Earth's annual motion around the Sun and its diurnal rotation on its axis. Such a statement is not quite correct. Copernicus convincingly showed that the assumption of such movements of the Earth eliminates the complications of the geocentric planetary theory with its numerous epicycles, explains the apparent rotation of the celestial sphere and provides the explanation of many astronomical phenomena from one simple principle.

There was a great deal of circumstantial evidence for the movements of the Earth, but as in a fair trial direct proofs were needed, and these were lacking.

In spite of the anonymous preface to his immortal book, Copernicus was convinced that he had discovered the true structure of the solar system and the movements of the Earth, but he had no direct proofs thereof, as he did not know of any facts, experiments, or phenomena, which could be explained only by the movements of the Earth and not in any other way. Even more, there was a simple geometrical fact directly opposed to the Earth's annual motion around the Sun and there was nothing that could directly prove its diurnal rotation. The dissenting fact was the absence of annual parallaxes of the stars, which were not as yet discovered nor could be detected by Copernicus himself. It was quite obvious from simple geometrical considerations that the orbital motion of the Earth must cause an apparent periodic oscillation of the stars. If in the motion of the planets epicycles appeared, that were a reflection of the Earth's orbital motion, the same, but on a much smaller scale, must appear in the position of the stars. As with the planets, the size of these oscillations depends on the distances of the respective bodies thus providing means for their determination. The stars did not show such oscillations and this was the strongest argument against the orbital motion of the Earth. This was not as naïve an objection as another which was propounded at his time, namely that a bird could not return to a nest, which had been carried away by the Earth during the bird's flight.

Copernicus, convinced of the correctness of the heliocentric system, could only retort that the stars do have parallaxes, but that they are too small to be observed. Such a statement was amazingly bold, as it implied that the distance of the stars is overwhelmingly great. If the precision of angular measurements at the time of Copernicus amounted to about 5 minutes of arc, the absence of noticeable stellar parallaxes meant that the stars were at least a thousand times more distant than the Sun and more than a hundred times farther than the most remote of the then known planets, Saturn, so that there was a tremendous gap between the planetary system and the surrounding celestial sphere. With increasing precision of measurements the dimensions of the stellar universe expanded and the lower limit of the distances of stars increased.

The absence of stellar parallaxes partly induced Tycho Brahe, the greatest observer of pre-telescopic times, to invent a clumsy and unfitted combination of the geocentric and heliocentric systems by keeping the Earth fixed in the center of the universe, and the planets revolving around the Sun, which moved around the Earth.

Soon the telescope was invented; in 1667 Jean Picard in France applied it to the measurement of angles, which increased the precision many tens of times. Nevertheless the stellar parallaxes remained unnoticeable. Further efforts of measurement were made not so much for a demonstration of the Earth's movement as for a determination of the distances of the stars. By this time the observations of Galileo, Kepler's laws of planetary movements, and the discovery of universal gravitation by Newton confirmed the movement of the Earth without any doubt. The existence of a celestial sphere was refuted and the stars were considered to be scattered at various distances in outer space.

Nevertheless continued attempts to measure stellar parallaxes were without any success. Such attempts were made by R. Hooke, J. Picard, G. D. Cassini, J. Flamsteed, J. Cassini, O. Roemer, and E. Manfredi, to mention only a few.

The English amateur astronomer, W. Molyneux, erected at his home at Kew near London an installation for observing the culminations of the zenith star γ Draconis situated not far from the pole of the ecliptic and therefore well suited for the detection of parallax. He invited James Bradley in Oxford to participate in the observations. In 1726 Bradley discovered a variation in the zenith distances at culmination with an annual period, amounting to 40 seconds of arc. In comparison with parallactic displacement, the star lagged behind by a quarter of a year.

It was probably easier to discover this strange displacement than to give an adequate explanation of it. It is said that Bradley, while walking on the bank of the Thames, noticed that the flags on the masts of vessels changed their direction each time when the ship altered its course, as the wind velocity was added to the speed of the vessel according to the parallelogram rule. It was already known that light propagated with a finite velocity, this having been discovered by Roemer half a century before. Bradley surmised that, like a flag in the wind, a ray of light coming from a star changes its direction when encountering the moving Earth, causing an apparent displacement of the star with the observed phase. This phenomenon called aberration of light proved the orbital motion of the Earth with no less certainty than if the annual parallax had been observed. The latter, however, still remained undiscovered.

Bradley, becoming Astronomer Royal in 1742, increased the precision of astronomical observations up to a few seconds of arc. The absence of noticeable parallaxes now meant that the stars were at least a hundred thousand times more distant than the Sun. In this case, photometric considerations showed that the luminosity of the stars is comparable to that of the Sun, i.e. that the Sun is one of innumerable stars differing from them only by its nearness to us.

The obvious smallness of stellar parallaxes made it quite hopeless to determine them by absolute methods or by comparing the coordinates of stars as determined from observations with meridian instruments. So determinations of relative parallaxes were attempted as had previously been suggested by Galileo. If there are two stars seen on the celestial sphere close together, but in reality at very different distances, then the parallax of the nearer star will be larger in the reverse ratio of their distances. In the course of one year it will appear that the angular distance between them changes by an amount equal to the difference of their parallaxes. This method has the advantage that only small angles have to be measured and this could be done by means of an ocular micrometer with much greater precision than a determination of the absolute coordinates of the stars.

By this time many double stars were discovered some of which consisted of stars of very different brightness. It was naturally assumed that the brighter star was much nearer to us so that such pairs presented convenient objects for the determination of relative parallaxes. A very assiduous observer of double stars at the end of the eighteenth century was William Herschel, who detected many of them and tried to use them for a determination of relative parallaxes. However, after 25 years of observing, in 1803 he came to the unexpected conclusion that most of the double stars were not perspective pairs seen only in one line of sight, but real binaries, stars close together and bound by their mutual attraction, undergoing orbital motion one around the other. Thus a new kind of celestial object was discovered, proving the universality of gravitation. Investigation of binary stars led to a determination of stellar masses, which were not very different from that of the Sun. In consequence a branch of stellar astronomy was inaugurated called astronomy of the invisible.

The discovery of double stars enriched science in an unexpected way, but they were quite useless for the purpose for which their observation was begun, namely relative parallax.

For 300 years the search for stellar parallaxes had continued, leading to several unexpected discoveries. After much futile seeking, towards the end of the 1830s the goal was reached. Nearly simultaneously in three different countries, for the first time in the history of astronomy, the annual parallaxes of three stars were observed and measured. The first to announce his result was F. W. Struve in 1837 before the completion of the building of the Pulkovo Observatory, then at Dorpat (now Tartu) in Estonia. Struve observed the parallax of the bright star of α Lyrae (Vega) with the largest instrument of that time, a 9-inch refractor by Fraunhofer from Munich. He measured micrometrically the angular distances of this star from two adjacent faint stars. Vega was chosen because its great brightness, together with a noticeable proper motion, made its proximity to the Sun very probable. F. W. Bessel, at Königsberg in Prussia, chose a rather faint double star 61 Cygni, guided only by its exceptionally large proper motion and measured the angular distance of its components also from two fainter stars with a heliometer—an instrument especially well suited for this purpose. Henderson at the observatory of Cape Town, South Africa, chose the bright double star α Centauri of the southern sky because of its brightness and large proper motion.

The largest parallax was obtained for α Centauri, reaching at the first determination nearly 1 second of arc, which corresponds to more than 200,000 astronomical units, i.e. 200,000 times the distance of the Earth from the Sun. The parallaxes of the two other stars were found to be 3–5 times smaller, corresponding to distances that many times larger. Subsequently the parallax of α Centauri was verified to be $0''.75$, so that its distance is 275,000 astronomical units or 4.3 light years and this, together with a faint accompanying star, remains the nearest star known at present. The outskirts of our Galaxy are several thousand times farther away. The dimensions of the realm of stars thus quite exceed all imagination.

The aberration of light and stellar parallaxes are direct proofs of the Earth's orbital movement around the Sun. Their values are interconnected, thus giving additional confirmation to the correctness of our interpretation.

The motion of the Earth must be taken into account in many astronomical investigations especially in our space age. If at the time of Galileo and Kepler its correctness was hardly to be doubted, the most convincing and direct proofs were given by the discovery of aberration and parallaxes of the stars.

So much as regards the Earth's orbital movement. Now let us proceed to proofs of its diurnal rotation.

Indirectly, the axial rotation of the Earth follows from the determination of the distances of the stars as evaluated by their parallaxes. Even from the point of view of ancient astronomy, assuming the stars fixed on a revolving celestial sphere with a radius corresponding to a parallax of 1 second of arc, the linear speed of diurnal rotation of a point on the equator of such a sphere would exceed 7500 times the velocity of light. Although this is obviously absurd, it is still not a direct proof of the Earth's rotation.

The polar flattening of the Earth, predicted by Newton and corroborated by measurements of arcs of the meridian in the first half of the eighteenth century, also showed the existence of rotation. This was especially obvious in comparison to the telescopically observed flattening of the rapidly revolving planets—Jupiter and Saturn.

However, again this is not a direct proof of the rotation of the Earth, as its flattening could be produced, although very improbably, by an appropriate distribution of matter inside the Earth.

In 1679 Newton pointed out a phenomenon, which if verified by observation would directly prove the Earth's rotation—this is the deviation of falling bodies to the east, being a consequence of the law of inertia. Newton's contemporary, Robert Hooke, who was "Curator of experiments of the Royal Society", tried to detect this but was unsuccessful. This is not astonishing owing to the smallness of height in his experiment—only 27 feet, the expected deviation amounting to less than 1/40th of an inch. Only a hundred years later, in 1791, the Italian D. Gugliemimi in Bologna dropped a lead ball from 240 feet and observed a deflection to the south-east of about two-thirds of an inch, which was a little larger than the calculated amount but within possible limits considering the crudeness of the experiment. Much better results were obtained by J. F. Benzenberg in 1802 at Hamburg from a church tower and in 1804 in a coal mine near Düsseldorf. In 1831 F. Reich in a pit near Freiberg obtained perfect agreement with the calculated value.

The Earth's rotation is demonstrated in a most convincing and evident way by a free-swinging pendulum, which has the property of preserving the plane of oscillation. The French physicist Leon Foucault installed such a pendulum 68 meters long in the dome of the Pantheon in Paris, and a multitude of spectators saw how the Earth turned under the pendulum that kept its oscillations in a permanent direction relative to outer space. Since then, this experiment has been demonstrated many times in different places, among them by Father Secchi, S.J., in Rome in the very church of St. Ignatius where in 1633 Galileo was compelled to deny that the Earth moves. There are also permanent installations of Foucault pendulums for continuous demonstration as, for instance, at St. Isaac's Cathedral in Leningrad.

Finally, it may be mentioned that the rotation of the Earth is utilized in the gyrocompass for a determination of the meridian direction in navigation. This instrument could not work in the absence of the Earth's rotation.

Such are the direct proofs of the movements of the Earth.

The Problem of Three Bodies: Copernicus and Modern Dynamical Astronomy

VICTOR SZEBEHELY

University of Texas at Austin

Introduction

Whittaker[1] calls the subject of this paper the "most celebrated of all dynamical problems", and indeed, with reason since it has been treated in several hundred publications by the most outstanding mathematicians during the past 200 years.

For detailed historical reviews Marcolongo's[2] and Lovett's[3] memoirs are recommended since in the present paper only the most significant contributions are reviewed in an undeservingly short manner.

The most important classical papers are by Whittaker,[1] Sundman,[4] Chazy,[5] Birkhoff,[6] and Siegel,[7] just to mention a few. During the last decade the analytical development followed Sundman and Chazy with Pollard, Saari, Lewis, Moser, Alekseev, Sitnikov, and Merman as the principal contributors. Papers by Agekian[8] and by Szebehely[9] in the last 5 years showed that the dynamical system under consideration is inherently unstable. The classical contributions are discussed in some detail with many references by Leimanis[10] and the more recent ones by Szebehely.[11]

The problem may be defined as follows. Three gravitationally interacting point masses are set in motion; find the subsequent behavior of the system. Since in the general problem the masses are of the same order of magnitude and the mutual distances are arbitrary, the assumptions made in the restricted problems of three bodies or in planetary and lunar theories are not valid. With arbitrary initial conditions the importance of special solutions, such as the Langrangian equilibrium configurations, are greatly reduced and relegated to classroom exercises. The few periodic orbits known in the general problem do not seem to be densely distributed, all seem to indicate instabilities and require specific initial conditions. The problem under the above general situation is then to predict the motion of the participating bodies from $t = 0$ to $t = \infty$.

The "solution" of the problem is a matter of definition. In the Newtonian sense the knowledge of the functions $\bar{r}_i = \bar{r}_i\,(t)$, representing the time-dependence of the position vectors of the bodies for all time *is* the solution. In the Laplacean or Poincaréan sense infinite series as the results of general perturbation methods are the solutions. In the Birkhoffian sense the qualitative behavior of the system is the solution. In the present paper this last approach to dynamics is followed and the qualitative behavior of the dynamical system is discussed. In this sense, then, we have the solution of the problem of three bodies. This qualitative solution has important physical meaning and it is valid for $t \to \infty$, for a great variety of initial conditions.

Qualitative Solution of the Problem

From Chazy's work[5] it is known that one of the pertinent dynamical parameters controlling the behavior of the system is the total energy $h = T - V$, where the kinetic energy is given by:

$$T = \tfrac{1}{2}\sum m_i\, \dot{\bar{r}}_i^{\,2}$$

and the negative potential energy or Poincaré's force function is:

$$V = \sum G\,\frac{m_i\, m_j}{r_{ij}},$$

where $1 \leq i < j \leq 3$.

Chazy classified the possible motions of the system for $h \gtreqless 0$ and found that for $h \geq 0$ at least two of the three distances increase to infinity as $t \to \infty$. In analogy to the problem of two bodies this result is not surprising. The difficulty is connected with the case when the total energy is negative. Chazy predicted for this case either a bounded motion, oscillatory motion, or the unbounded increase of two of the mutual distances to infinity. This classification has been refined recently and a description, physically more meaningful has been proposed by Szebehely.[12] Here only the case of $h < 0$ will be discussed in detail. The final result of recent investigations may be announced as follows. As $t \to \infty$, the motion asymptotically approaches the separation of the three bodies into a binary and an escaping body. For arbitrary initial conditions the bodies may perform, for an initial period of time bounded motion, called *interplay*, or one of the bodies may be *ejected* on a perturbed elliptic orbit relative to the two other bodies. Sooner or later, however, such bounded configurations cease to exist and one of the bodies *escapes* on a perturbed hyperbolic orbit relative to the two other bodies. These two bodies form a binary, the orbit of which asymptotically approaches an ellipse as the escaper departs to infinity.

Initial conditions belonging to sets of measure zero are excluded such as equilibrium solutions, periodic orbits, or rotational motions. The rotational configuration, properly called *revolution*, consists of a binary surrounded by the orbit of the third body. This motion requires special initial conditions and it occurs only for $h < 0$. Its stability depends on the ratio between the separation of the third body from the binary as shown by Harrington.[13] When this ratio becomes approximately unity the revolution becomes an interplay and an escape may be expected sooner or later.

The phenomenology of the eventual escape-mechanism is as follows. Negative total energy allows, in fact enhances, triple close approaches. When this happens the value of V becomes large and according to the Lagrange–Jacobi equation, the second time-derivative of the moment of inertia (I) becomes positive, since:

$$\ddot{I} = 2\,(V+2h).$$

Note that if the total energy is not negative ($h \geq 0$), then $\ddot{I} \geq 0$ always, but if $h < 0$, a close approach is needed to have $\ddot{I} > 0$. The condition of $\ddot{I} > 0$ means that the curve $I(t)$ is convex. After the minimum moment of inertia has been obtained, V decreases as the bodies separate and $\ddot{I}$ may become zero or negative. Escape requires that $\ddot{I} > 0$ as $t \to \infty$ and this may be obtained if after $I_{\min}$ the bodies are so arranged that $V+2\,h > 0$ or since $h < 0$, $V > 2|h|$. The quantity V may be maintained at a high value after the triple close approach if one of the three mutual distances stays small. In this way the other two distances can increase to infinity and $\ddot{I}$ still remains positive. The process may be described by requiring that:

$$V = G\left(\frac{m_1\,m_2}{r_{12}}+\frac{m_2\,m_3}{r_{23}}+\frac{m_3\,m_1}{r_{31}}\right) > 2|h|.$$

Let now r_{12} be the smallest of the three distances. Then:

$$2|h| < V < G\frac{A}{r_{12}},$$

where $A = m_1\,m_2+m_2\,m_3+m_3\,m_1$, or:

$$r_{12} < \frac{GA}{2|h|},$$

i.e. as long as the smallest distance, r_{12}, is below the value of $GA/2|h|$ the expansion will occur with $\ddot{I} > 0$ and one body escapes. Equating r_{12} with the pericenter of the binary formed by m_1 and m_2, we have:

$$\frac{AG}{2|h|} \geq a(1-e).$$

Given the masses and the total energy, the properties of the binary formed determine whether escape occurs or not. If a close binary with high eccentricity is formed, escape is more probable than when a circular binary with large semi-major axis is formed. It also follows that if the total energy is close to zero, the probability of escape is high, as expected.

The phenomenology may be also followed by an approximate energy consideration. The total energy h may be written, after a binary is formed, as $h = E_b + E_e$, where E_b is the energy stored in the binary and E_e is the ejection or escape energy of the third body. If $h \geq 0$, we have $E_b + E_e \geq 0$ or, since the energy stored in the binary is always negative, $E_e \geq |E_b|$. This means escape, since $E_e > 0$. Therefore, for positive or zero total energy a binary formation always results in an escape.

If $h < 0$, we have $|E_b| - |h| = E_e$, therefore, the formation of a binary is a necessary but not a sufficient condition for escape when $h < 0$. The energy stored in the binary is $E_b = -Gm_1 m_2/2a$ and, therefore, positive escape energy requires that:

$$G\frac{m_1 m_2}{2a} > |h| \quad \text{or} \quad \frac{Gm_1 m_2}{2|h|} > a.$$

Once again we see that the formation of a close binary is necessary for escape. Note that the limit of a for binary formation is higher when the Lagrange–Jacobi equation is used, since:

$$\frac{A}{2|h|\,(1-e)} > \frac{m_1 m_2}{2|h|}.$$

Astronomical Consequences

Regarding isolated triple stellar systems the result of the Laplacean instability indicates that triple configurations do not exist as $t \to \infty$. Triple systems exist only for negative total energy and only as temporary configurations. The process of separation takes over faster for $h > 0$ than for $h < 0$. The formation of a binary always gives an escape for $h \geq 0$ but only close binaries result in escape when $h < 0$. If $h < 0$ the formation of a binary with large semi-major axis results in an elliptic ejection of the third body which will rejoin the binary for further interplay. Revolution is the only stable configuration with sufficient separation of the binary from the third member. If the separation is sufficiently large, the third body may be disturbed by other stars and it will leave the binary. If the separation is too small, the system becomes unstable. The presently known triple stellar configurations correspond to this type of motion. Their long-time stability is not evident. The time of disruption is of the order of 10^9 years as an average value for arbitrary initial conditions.

The escape is the consequence of three-body effects and it always follows a triple close approach. Computational difficulties may be expected when triple close approaches occur since triple collisions represent non-regularizable singularities. The phenomenon of triple close approach, responsible for the numerical difficulties as well as for the dominant behavior of the system (escape), seems to indicate, once again, how masterfully nature masks its laws. The role of angular momentum is to delay collapse and, consequently, to delay escape.

The characteristics of the binaries formed by dynamical effects vary greatly and depend mainly on the total energy of the original triple system.

Saari's[14] results indicate that initial conditions which result in binary or triple collisions form a set of measure zero. Collisions here are to be understood in the mathematical point-mass sense. Numerical results indicate that the smallest distances are large enough not to be concerned with tidal effects but small enough to demand the use of regularization techniques in the numerical integrations.

Usually, but not always, the body with the smallest mass escapes. The stellar dynamics of triple configurations is considerably different from the celestial mechanics of the triple configurations used in planetary and lunar theories. One major difference is the denseness of periodic orbits in the restricted problem and their lack in the general problem, where the asymptotic escape orbits are dense. The techniques of celestial mechanics, based on periodic solutions, therefore, must be replaced by techniques which use asymptotic escape orbits.

A triple stellar system when placed into the field of a cluster should show similar behavior since the initial mechanism is the triple close approach and the subsequent escape. The continuation of the numerical integration of the motion of a triplet after a sufficiently close triple approach is not necessary if the isolated behavior is studied since escape is predictable in this case analytically. Such continuation "through" the triple close approach is mandatory if the escaper and the binary are members of a cluster. Analytical predictions are available which offer estimates of the lower limit of the closest distance during a triple close approach using the conditions at earlier times. Using such estimates the accuracy requirements of the numerical integration may be estimated.

Acknowledgments

This paper was prepared during the author's sabbatical sponsored partly by the University of Texas Research Institute and by the National Research Council of the U.S. National Academy of Sciences during his tenure at the Manned Spacecraft Center of NASA in Houston.

References

1. E. T. WHITTAKER, *A Treatise on the Analytical Dynamics of Particles and Rigid Bodies*, Cambridge University Press, 1904.
2. R. MARCOLONGO, *Il Problema dei Tre Corpi*, Hoepli Publ., Milano, 1919.
3. E. O. LOVETT, *Quart. J. Math.* **42**, 252, (1911).
4. K. F. SUNDMAN, *Acta Math.* **36**, 105 (1912).
5. J. CHAZY, *Bull. Astron.* **35**, 321 (1918).
6. G. D. BIRKHOFF, *Dynamical Systems*, Am. Math. Soc. Publ., Providence, R.I., 1927.
7. C. SIEGEL, *Ann. Math.* (2), **42**, 127 (1941).
8. T. A. AGEKIAN and ZH. P. ANOSOVA, *Astron. ZH.* **44**, 1261 (1967).
9. V. SZEBEHELY, *Proc. Nat. Acad. Sci. USA*, 58, 60 (1967).
10. E. LEIMANIS and N. MINORSKY, *Dynamics and Nonlinear Mechanics*, Wiley Publ., N.Y., 1958.
11. V. SZEBEHELY, "The problem of three bodies", in *Recent Advances in Dynamical Astronomy*—Proceedings of the NATO Advanced Study Institute in Dynamical Astronomy, Cortina d'Ampezzo, 1972, Publ. D. Reidel Co, Holland, 1973.
12. V. SZEBEHELY, *Cel. Mech.* **4**, 116 (1971).
13. R. S. HARRINGTON, *Astron. J.* **73**, 190 (1968).
14. D. SAARI, *Cel. Mech.* **7**, 2 (1973).

Copernicanism in Bruno, Kepler, and Galileo

STILLMAN DRAKE

University of Toronto, Toronto, Canada

THE names of Giordano Bruno, Johann Kepler, and Galileo Galilei would undoubtedly be remembered today even if Copernicus had not published his *De revolutionibus*. Yet it happens that those three men are remembered as the foremost Copernicans of the century following the publication of that astonishingly germinal book. Both these facts deserve a few minutes of our attention on this occasion, when we meet to pay homage to Copernicus rather than to celebrate his followers.

If anyone doubts that Bruno, Kepler, and Galileo would have made their marks in the intellectual history of the western world without drawing on the work of Copernicus, that is because their chief works seem to center on the new astronomy, and we do not ordinarily pay much attention to their other works. In the case of Bruno, Miss Frances Yates has made it more than clear that his role in transmitting and developing certain mystic traditions that have been nearly as important as science itself in shaping our intellectual past assured him a permanent place in the history of ideas. As to Kepler, I shall mention as examples of his non-Copernican contributions only his classic work on optics, published in 1604, and his discussion of the measurement of volumes of solids of revolution published in 1615. And with regard to Galileo, it is clear that his studies of motion, mechanics and the strength of materials far transcend in scientific importance anything he had to say about astronomy.

That the same three men are in fact best known to us at present as vigorous supporters of Copernicus will, I think, be questioned by no one. But there is one curious aspect of this which tends to escape notice, and it is one on which I intend to dwell. It is that each of these three men altered the system of Copernicus in his own way. Copernicanism in Bruno has in fact very little to do with *De revolutionibus*; that great astronomical work merely afforded to Bruno a vital link in his vast and eclectic philosophy, with parts of which Copernicus himself would never have agreed. Copernicanism in Bruno is vanishingly small, while the Copernicanism of Bruno is its infinite extension. The case is very different with Kepler. Virtually everything in *De revolutionibus* is also dealt with by Kepler, but nearly everything essential to it becomes modified; and where Kepler extended the original work, his extensions were more physical than metaphysical, as they had been with Bruno. Finally, Galileo's Copernicanism differs from that of *De revolutionibus*, as do Bruno's and Kepler's; but Galileo's presentation, far from being an extension of the work of Copernicus, imposed a drastic restriction on it, putting it in a form that Copernicus had long ago realized to be untenable, and had abandoned.

And so, in capsule form and therefore somewhat distorted, I shall say that Bruno gives us an extended Copernicanism, Kepler a modified Copernicanism, and Galileo a restricted Copernicanism. That the story is not even as clear and straightforward as that is well known to most of you, and will become evident as we proceed. For the points of contact and the points of departure between the ideas of all four thinkers are myriad.

What has been said thus far is intended to throw light on a situation remarked by Professor Thomas Kuhn in his book, *The Copernican Revolution*. Speaking, I believe, for most historians of ideas today, he said of *De revolutionibus* that it

> must be a constant puzzle and paradox; for measured in terms of its *consequences*, it is a relatively staid, sober, and unrevolutionary work. Most of the essential elements by which we know the Copernican revolution—easy and accurate computations of planetary positions, the abolition of epicycles and eccentrics, the dissolution of the [solid] spheres, [consideration of] the sun as a star, the infinite expansion of the universe—these, and many others, are not found anywhere in Copernicus' work.[1]

If I may be permitted to use a word that Professor Kuhn later made famous, what seemed to him a paradox of the Copernican revolution may now be seen alternatively as a paradigm of true scientific revolutions. It seems to be the rule, rather than the exception, that the most enduring and wide-ranging alterations in patterns of scientific thought have been based on relatively staid, sober and unrevolutionary works, devoid of attacks or polemics and concerned with harmonizing known facts. Darwin's *Origin of Species* was such a work, and so was Einstein's *Zur Elektrodynamik bewegter Körper*; yet both ushered in vast alterations of scientific thought. And as with the work of Copernicus, the excited attention they attracted was dependent largely on certain embellishments added to them by zealous followers or bitter opponents. Thus Bruno's romantic classification of our Sun as a star, and

[1] T. Kuhn, *The Copernican Revolution*, p. 134, Cambridge, Mass., 1957.

his mystical assertion that the Universe is infinite, with infinitely many inhabited worlds in it, were later to be matched by the horrifying description of man as a direct descendant of the ape, and the despairing idea that all human values were put in doubt by relativity—declarations not to be found in Darwin's book or Einstein's paper, but regarded as necessary implications thereof by other men.

The attraction of Kuhn's paradox arises not from the sobriety of *De revolutionibus*, but from our indiscriminate uses of the word "revolutionary". Experience teaches us that the durability of revolutions affecting social institutions is measured by the power of slogans to continue to excite emotion, rather than by their degree of rationality. Experience should likewise teach us that the durability of a scientific revolution is little affected by any attendant romance or despair; rather it is measured by the very sobriety of the work that led to the revolution. Had Copernicus made his system depend on, or even imply, an infinite universe, it would not have appealed to either Kepler or Galileo, and might well have provoked their opposition.

As it was, the principal opposition expected by Copernicus was identified in his letter of dedication: "It may happen that certain exegetes, ignorant of mathematics, may feel privileged to pronounce judgement on my work by reasons of this or that scriptural passage twisted to their purpose. Should any such persons venture to criticize my meaning, I take no account of them; I consider their judgment rash, and I despise it."[2] Copernicus expected theological attacks, but was confident that mathematicians would approve his work. It is ironical that the first theologian to attack it, Martin Luther, did so more on grounds of common sense than of scripture, while the first vigorous attack incorporating biblical grounds came from the rival astronomer and mathematician, Tycho Brahe.

The first articulate literary and philosophical enthusiast for Copernicanism, on the other hand, seems to have been the theologically trained ex-Dominican Giordano Bruno, with whom the wide philosophical and theological implications of the new astronomy—or I should say, its supposed implications—took the center of the stage (Fig. 63 on p. 191). The only truly Copernican elements in Bruno, as I see them, were less those of heliocentrism than those of the uniformity of nature and the relativity of motion; and, of course, the same demand for freedom in philosophizing untrammeled by irrelevant criticism that I have already cited from Copernicus' preface. The uniformity of nature, so essential to Bruno, appeared in *De revolutionibus* only incidentally, in some rather brief comments reconciling physics with the rotation of the Earth and accounting for the spherical shape of the heavenly bodies and the earth. Relativity of motion was utilized by Copernicus only in the sense of optical relativity, a necessary datum of all astronomers, with the addition of the famed lines of Virgil concerning ship and shore in order to emphasize the situation of an observer on a moving earth.

[2] N. Copernicus, *Dedication* to *De revolutionibus orbium coelestium* (1543). Cf. Kuhn, p. 142, for a slightly different translation.

The wide-ranging and eclectic philosopher Bruno promptly extended those restricted principles of Copernicus into absolutes, so to speak. In saying "promptly", I ought to state when and how, in my opinion, Bruno first became acquainted with and an enthusiast for the Copernican theory. Born at Nola, near Naples, in 1548, Bruno entered the Dominican order at an early age, but after a time he found himself assailed by doubts and in conflict with the authorities. From 1576 on, he was a wanderer, at first in Italy and then in Switzerland and France. He first gained fame for his phenomenal memory, and for a system of memory-training that interested the King of France. Bruno's first book, published at Paris in 1582, dealt with memory in a framework of magic, and on his arrival in England in 1583 he continued with a second work on the same subject. There is no suggestion that Copernicanism had yet attracted him. But in the next year he advocated the Copernican system strongly, at Oxford and in London, and from that time on, it entered deeply into his main philosophical works. The way in which it appears there, as well as this timing, suggests that Bruno's first, and perhaps his only, acquaintance with the doctrine of Copernicus came through a certain British Copernican work.

It is well known that Copernicus had in no way suggested the idea of an infinite Universe, which was to become Bruno's theme. For the purposes of the Copernican astronomy, it was sufficient to postulate that the sphere of the fixed stars was remote enough to mask any parallactic effect of the Earth's annual motion. Indeed, neither Kepler, who admired Bruno and regarded him as a Copernican, nor Galileo, whose writings suggest that he was not unacquainted with Bruno's, ever for a moment accepted his notion of an infinite Universe. It is clear, then, that Bruno's conception was not a normal and standard extrapolation from anything that Copernicus himself had written.

Now, it is also well known, as a result of researches by Francis Johnson and Sanford Larkey in the early 1930s, that the English Copernican Thomas Digges had published in 1576 a posthumous revised edition of his father's *Prognostication Everlasting*, and that in this edition, there appeared a translation of the opening parts of *De revolutionibus* together with a modified Copernican diagram and a clear and unequivocal statement asserting the infinite expanse of the fixed stars. Whether Thomas, or Leonard, Digges was originally responsible for taking this momentous step is not certain, but the reasoning behind it is quite evident. If the diurnal motion of the fixed stars is removed, and replaced by a rotation of the observer on Earth, then there is no longer any reason to confine those stars in a spherical shell, no matter how large; and in scientific thought, a restriction that is not necessary is quite meaningless. Hence the fixed stars, being without motion, could expand to infinity without doing violence to conventional physics, though this did contradict conventional astronomy.

It was Digges' work that I believe came into Bruno's hands in 1583, 7 years after it was published, and shortly after his arrival in England. The infinite Universe that

Bruno associated with Copernicanism immediately became the keystone of his arch of eclectic philosophy in which nothing was excluded; for Bruno wrote:

> There is no kind of philosophy that has been ordered by tempered feeling which does not contain something good of its own that is not contained in the others . . . I praise you if just as you are not as ordinary in philosophy as Aristotle, you are also not as injurious to others and as ambitious as he, for he deprecates the opinions of others, together with their methods of philosophizing. . . . The Epicureans said many good things, though they never rose above material considerations. Heraclitus knew much that was excellent, though he never went beyond the spirit. Anaxagoras did not fail to profit from nature, for he not only went into this but beyond it, and perhaps above it, to know that intellect which is called God by Socrates, Plato, Trismegistus, and our theologians.[3]

In Bruno's vast tolerance I perceive a reaction against the unprecedented fragmentation that characterized the times in which he lived. It was not only religion that had been split apart about the time he was born, but, as compared with the Middle Ages, chaos seemed to reign in the politics, the literature, the art, music, and philosophy of the late Renaissance. The universities no longer enjoyed a monopoly on learning, and Latin was losing its unifying hold on learned men. The spectacle of division among his fellows could not but disturb Bruno's sensitive soul. Now, to Bruno, the concept of infinity implied a kind of superlative unity, and it was that doctrine that he preached from the time of his sojourn in England to the end of his life.

It was the British Copernicanism of Digges, I believe, that finally crystalized in Bruno's mind that mass of paradoxical and visionary ideas that his vast memory had been accumulating from philosophers, mathematicians and mystics during his years of wandering and debate. I do not mean that Bruno turned to Copernicanism as a lone beacon of unification in a fragmented age, but I do think that he saw the new astronomy as a parable for philosophers, one in which a single unifying principle had brought order into the welter of unrelated orbs of the Ptolemaic astronomy. In a sense, that unification was the principal virtue that Copernicus himself had seen in his system.

Treating the solar system as a parable or metaphor, Bruno declared that every fixed star was a sun like our Sun; and since there was no reason to suppose a limit to their number, he deduced that they must be infinite in number and extent. An infinity of inhabited worlds circulated around his newly discovered infinity of suns. All motion was relative, not only optically but physically, and some of Bruno's illustrative examples of this reappear in Galileo's physics. On the analogy of infinite worlds, ever expanding outward, Bruno envisioned ever-smaller worlds of ever-moving atoms. The whole was then animated, in that part of Bruno's philosophy that deals with the world-soul and with the monad; his psychic atoms draw energy from that single universal center which unites not only all mankind, but all things, infinite

[3] G. Bruno, *De la causa, principio, et uno; Opere*, ed. A. Wagner, vol. 1, pp. 258, 260 (Leipzig, 1830).

in number. Bruno did not manage to present acceptably that subtle linkage that does indeed exist between the concepts of infinity and of unity, but his mathematical gifts were later to command the sympathy and respect of such a peerless mathematician as Leibnitz.

All this, however, did his cause little good at the time, even in England. Most men failed to see the logical linkage between infinity and unity, and instead saw the psychological connection between infinity and chaos, of which they had had enough. The general response was epitomized in those celebrated lines of John Donne, in which the "new philosophy" called all in doubt and shattered all coherence. This new philosophy can hardly have been anything but that of Bruno, who had crumbled the world into atomies before Gassendi did, and whose relativity had denied a fixed place even to the Sun itself. This was not really Copernicanism; but in Bruno's mind it was, and it came to be linked with Copernicus in the literary world through Bruno's writings. Along the lines of this reconstruction, we can make some sense of the romantic nineteenth-century notion that Bruno was a martyr to Copernicanism. For Bruno's final program was to reunite philosophy and science in a single world religion, a program that went up in smoke with his frail body in the last year of the sixteenth century. In its place, a very different scheme of things was to develop in the seventeenth century, and one which was no less linked with Copernicanism, as we shall presently see. But here, in his own words, is an eloquent summary of Bruno's vision of infinite unity, inspired by British Copernicanism and utterly remote from the staid, sober, unrevolutionary *De revolutionibus*:

> Therefore the universe is one, infinite, and immovable. It is one, I say, in absolute possibility; one in act; one in form or soul; *the* one, greatest and best, which can never be comprehended, and is therefore infinite and boundless, and likewise infinite and unbounded, and consequently immovable. It does not move in position, because there is nothing outside it whither it is carried, inasmuch as it is all. . . . It is not matter, because it has no shape and cannot be shaped, not being bounded or capable of bounds. It is not form, because it does not give form to anything else, inasmuch as it is all, and one, and *the* universe. Being one and the same, it does not have this being and that being, and hence does not have this part and that part; and since it does not have parts, it is not compounded. This is [to have] bounds of a sort that are not bounds, and form that is not form, and matter that is not matter, and soul that is not soul, because it is *all*, indifferently; and therefore it is one, the universe is *one*.[4]

For such ideas, Bruno was burned at the stake; and though they do not include anything we should recognize as Copernicanism, I believe they might never have emerged in Bruno's mind had it not been for his acquaintance with Copernicus through the work of Digges.

Four years before Bruno's death, Johann Kepler published a book that has a certain affinity with the mysticism of Bruno, whom Kepler greatly admired. It was called the *Mysterium cosmographicum*, or rather, the prelude to the cosmographical mystery that Kepler was determined to penetrate (Fig. 64 on p. 191). This book

[4] G. Bruno, *op. cit.*, p. 280.

was as remote from anything dreamt of by Copernicus as was the infinite unity of Bruno. Kepler sought in it to determine by purely mathematical considerations the reason for the number of the planets and for their placements with respect to the Sun. The project was not crowned with success, but it set the keynote of all Kepler's work in astronomy, which was later to achieve the most overwhelming success. That keynote was not the *unity* of the Universe, as with Bruno, but rather with the *harmony* of the Universe, which is quite a different thing. There is in a sense no harmony in unity, but only monotony, even though musicians of the time counted unison as a form of concord. Where Bruno denied that the Universe had parts, Kepler numbered its parts, put them in order, and harmonized them geometrically. And as Copernicus had taken the first step toward unity when he reduced a great many seemingly diverse planetary motions to a single pattern, so he had also, by that same reduction, taken the first step toward harmony. The difference was not in what Copernicus had done, but in how that was seen by Bruno and by Kepler. It is perhaps unnecessary to add that it was seen still differently by Galileo.

The literalness with which Kepler took the notion of harmony as applied to the heavens leaves no doubt whatever about the source of his original inspiration; it was the classical Pythagorean "music of the spheres". But during the next few years, Kepler's Copernican convictions found another quite different basis. This resulted from his acquaintance with Tycho Brahe, and centered on the use of the extremely accurate observations of the Danish astronomer. The undoubted reliability of those data led Kepler to concern himself with the precise computation of orbits, though not forsaking their Pythagorean mysteries; and in his great *Astronomia Nova* of 1609 there is little trace of his earlier preoccupation with mathematical oddities as the source and cause of the distribution of the heavenly bodies. In their place is the systematic and painstaking pursuit of structural realities through careful and laborious computation from empirical data. This procedure seems to me to have been more truly in line with the true example of Copernicus; and, as with Copernicus, a bold new theory emerged from Kepler's meticulous respect for observational facts. With Copernicus, what had thus emerged was the heliostatic idea, breaking with ancient tradition. With Kepler, what emerged was an equally sharp break with tradition and convention in the discarding of circular orbits for elliptical paths, and of uniform for non-uniform motions. Not only that, but the concept of "center" itself was altered.

Should the new Keplerian astronomy of 1609 be called Copernican, or not? Kepler himself had no doubt whatever that it should, and that in itself is suprising when we consider the custom of the period, and the general habit of all discoverers. Tycho Brahe was inordinately proud of the system which he outlined, but never worked out in full detail, and he fought vigorously against any challenge to his priority with respect to it. To our eyes, it is merely a derivative of the Copernican system, achieved by the arbitrary choice of the Earth as the resting body in a set of relatively moving objects. To the extent that Tycho's system could have been completely worked out, in any intelligible sense of those words, it would have been no more than Copernicanism with a change of coordinates. Kepler himself showed this geometrically a number of times, in loyal defence of Tycho against his foes. Yet Tycho insisted that the system be regarded as truly original with him, and it still bears his name, even though there is now strong evidence that it had had a forerunner in ancient Greece. In contrast, Kepler's astronomy was truly novel in two important respects never anticipated by anyone—his elliptical orbits, and the irregular speeds therein. Kepler, however, did not even suggest, let alone insist upon, a renaming of the system, and he continued to declare himself a Copernican even after drastically remodeling the Copernican system in essential, and not merely in trivial respects.

There were excellent reasons for Kepler's continuing to speak of his system as Copernican. One reason was the single-minded concern of both astronomers with the goodness of fit of observational data to the heliostatic assumption, so that Kepler's achievement seemed to him a carrying out of the same program with only an increase in accuracy. Another area of kinship lay in a certain attitude toward the sun itself, expressed by Copernicus in these words:

> In the middle of all sits the sun enthroned. In this most beautiful temple, could we place that luminary in any better position from which it could illuminate the whole at once? It is rightly called the Lamp, the mind, the ruler of the universe; Hermes Trismegistus calls the sun the visible god, and Sophocles has Elektra call it the all-seeing. So the sun sits as on a royal throne ruling his children, the planets, which circle round him.[5]

Just what this isolated literary passage signifies, standing alone in the vast technical bulk of *De revolutionibus*, may reasonably be questioned. Most historians today pretend that it shows the innermost workings of the mentality of Copernicus. I cannot prove that Copernicus was not led to his astronomical system by contemplating the Platonic philosophy, the mysticism of the legendary Hermes, and the casual opinion of a character in Greek drama. I think it more likely that Copernicus became an enthusiast for Hermes Trismegistus only after he had arrived at his heliocentric astronomy, and that he was led to it not by respect for mystical opinion, but by a wish to improve the certainty of the calendar. In either case, though, this sort of Sun-worship is most certainly identifiable as an element of Copernicanism in the work of Kepler, who wrote: "Called king of the planets for his motion, heart of the world for his power, its eye for his beauty . . . by the highest right we return to the sun, who alone appears, by virtue of his dignity and power, suited for this motive duty and worthy to become the home of God himself, not to say the First Mover."[6]

This very element of Sun-worship in Copernicanism had been instrumental in Kepler's consuming desire to establish a physics of the heavens. Asserted on the

[5] N. Copernicus, *De revolutionibus*, Bk. I, ch. 10. Cited by Kuhn, p. 179.

[6] J. Kepler in a fragment cited by E. A. Burtt, *Metaphysical Foundations of Modern Science*, p. 59, Anchor Book ed., N.Y., 1954. Cf. J. Kepler, *Prodromus* (1596), ch. 20.

title-page of his *Astronomia Nova* and exemplified in that, as in all his later Copernican works, this objective had been foreshadowed in his first book, which concerned itself with physical causes as well as mathematical reasoning. For it did not seem to Kepler, any more than to Galileo, that a vacant mathematical point, a mere nothing, could, by virtue of its being the center of the Universe, exert control over physical bodies of any kind, and especially over the vast celestial bodies with which astronomy is concerned.

Copernicanism in Kepler was guided on the one hand by a real concern for physics, and on the other hand by an equally real demand for mathematical precision in dealing with the architecture of the Universe. Yet both these guiding principles differ somewhat from their counterparts in the work of Copernicus himself. The founder of the new system was content that it be free from physical incongruities. Thus it was only in the introductory sections of *De revolutionibus* that questions which may properly be called physical were brought up at all, and they were examined only to the extent necessary to reconcile the assumed motions with the traditional concepts taken from Aristotelian physics. Likewise, it was not so much the demand for mathematical precision as the desire for mathematical elegance that is reflected in the work of Copernicus.

Lest my intention here be misunderstood, I shall amplify these remarks in this way. It seems to me that the willingness of Copernicus to displace the real Sun somewhat from the center of every planetary orbit reflects his acceptance of the traditional role of the astronomer, in which the mathematics of description was his primary task, and all physical questions were left to the philosopher. It was Kepler rather than Copernicus who first refused to recognize as legitimate any such separation of tasks. And if Kepler did not succeed in unifying astronomy and physics, he did succeed in finding the first unifying laws that were destined in time to make possible the completion of the program he envisioned—a program considerably more ambitious than that which Copernicus had had in mind, though in no way inconsistent with it. In this regard, Bruno and Kepler both wished to expand the Copernican theory. Each went beyond the scope and intent of the original scheme, Bruno in the direction of metaphysics, and Kepler in the direction of physics.

When I say that Kepler did not succeed in unifying astronomy and physics, I refer to the fact that this is something that he consciously attempted, and to the fact that we have since been compelled to reject the physical part of his theory. I do not think it is a good practice to say that an early scientist "failed" to do things done later if he never tried to do them; it is sufficient to say that he did not do those things. Thus it would be inappropriate, in my view, to say that Copernicus failed to introduce non-circular paths, or that Galileo failed to explain what held the planets in their orbits, since those are things they did not attempt at all. Kepler did attempt to determine the physical forces at work in the heavens, and in this he failed. But what a glorious failure it was! Had he merely applied his own photometric law to the Sun's force in turning the planets, he would have had the inverse-square law of Newton in his grasp. And even the general idea of a mutual gravitation of bodies was applied by him in an attempted explanation of the tides.

The roots of Kepler's failure to unify astronomy and physics were themselves Copernican in character. For just as Copernicus was content to show how certain physical questions might be handled within an accepted framework, so Kepler's attempt to enlarge that program was doomed to failure only because the old physics was inadequate to meet the needs of the new astronomy. Parallel instances have occurred in later science; and generally speaking, the alteration of the basis of one science in order to accommodate the requirements of another is justified only after a serious attempt has been conscientiously made to preserve both, and that task has found to be impracticable. The credit for such an attempt belongs to Kepler.

The actual emergence of a new physics capable of accommodating the new astronomy, however, begins with Galileo, and it seems to have been largely coincidental rather than deliberate (Figs. 65 and 66 on p. 192). The way of writing history of science as a history of ideas, rather than as an account of the men who contributed to their development, has concealed the accidental character of this historical fact. Because Galileo laid the basis of the required new physics, and because Galileo also took part in the battle for the Copernican astronomy, it is supposed by most historians that the original inspiration for Galilean physics was that he felt the need to reconcile observed terrestrial phenomena with a moving Earth. The assumption has no basis in documents known to me, and it is in sharp disagreement with a great many such documents. If we are to rely on the record, it is clear that Galileo's first, and for a long time his only, interest in the Copernican astronomy was that it provided him with a supposed mechanical explanation of the tides. That does not represent a need to reconcile terrestrial phenomena with a moving earth; on the contrary, it appeals to a moving Earth to explain an observed terrestrial phenomenon. Perhaps this sounds like pretty much the same thing, but it is not so at all. What historians mean is that the new astronomy could not be accepted until a terrestrial physics appeared in which it could be explained why a body falling from a tower did not strike the ground somewhat to the west of its base. That is quite true. But it is not true, as they suppose, that Galileo became interested in motion through any attempt to provide an explanation for such facts. He did eventually provide the explanation, but that was a by-product of long investigations that had had a totally different origin. And Galileo's theory of the tides was formulated about 1595, which was many years before he offered any explanation for the vertical paths of bodies falling to a moving Earth.

Galileo's preference for the Copernican system first became of record in two letters written in 1597, and his first published mention of it did not occur until 1610. Had he died in 1609, and had his notes and correspondence been preserved only through that year to the extent that they do survive today, we should find among them virtually all of his terrestrial physics, including the laws of falling bodies, the ideas of conservation of motion and composition of motions, and the derivation of the parabolic trajectory. But we should find no significant mention of Copernicus or of the Copernican system beyond the two letters of 1597, and an oblique reference in a syllabus prepared for his students around 1600. Galileo was in fact already engaged in the composition of a book on terrestrial physics in mid-1609, when he first heard of the invention of the telescope. What made him a convinced and enthusiastic Copernican were his own telescopic observations early in 1610. Had it not been for the telescope, it is doubtful that Galileo would ever have become closely identified historically with Copernicanism. In any event, it is pretty certain that Galileo's fundamental physics neither resulted from, nor led him to, Copernicanism; nor did he add significantly to his knowledge of terrestrial physics after 1610. And yet it was precisely because of Galileo's terrestrial physics that he became an effective propagandist for the new astronomy.

This may sound paradoxical, but there is nothing inherently implausible about it; it merely offends our sense of causal linkages in history. Terrestrial physics and celestial physics are now so closely linked that we can hardly think of them as ever having been separate; yet their separation had been the whole basis of Aristotle's natural philosophy, which no one had successfully challenged in 2000 years. It was really Kepler who first set forth a serious challenge, as mentioned earlier, and he introduced physics into astronomy against the warnings of his teacher, and despite a very old tradition among astronomers. Thus there was nothing remarkable about Galileo's having given years of research to terrestrial physics without a thought—or at any rate without a written note—concerning a possible application of it to the heavens.

Indeed, Galileo never did attempt to explain the motions of the planets physically, even in his pro-Copernican works. His interest in the physics of the heavens had to do not with the planets, but with lunar mountains and sun spots. And if we look closely at the *Letter to Christina* and the *Dialogue*, we shall find that those works are not so much pro-Copernican as they are anti-anti-Copernican. There is a good reason for this. The *Letter to Christina* was written in 1615, and its purpose was to prevent the official banning of the Copernican theory by the Church. Hence it was designed mainly to counter theological opposition, rather than to support the new astronomy. This battle was lost in 1616, when an official edict forbade the supporting of Copernican ideas in print, and even prevented refutation of scriptural arguments against Copernicus. But it did not specifically forbid the refutation of any astronomical, mathematical or physical arguments, and hence the *Dialogue* was necessarily composed mainly along those lines.

Galileo's Copernicanism in the *Dialogue*, as mentioned earlier, is not an extended version as with Bruno, nor is it the correctly modified version of Kepler; it is modified, but very incorrectly, in a direction that Copernicus had already rejected. For Galileo, in presenting the new system to his readers, placed the Sun at the center of all the orbits, which he represented as perfectly circular. This may mean that Galileo was an incredibly poor and ill-informed astronomer, since no set of astronomical tables ever compiled would agree with such an arrangement. But it seems more likely that he presented it in this way to simplify the essential ideas for his readers, his book being especially designed for laymen. Nor was such a presentation temperamentally unsuited to Galileo, whose watchword in science was simplicity in nature, just as Bruno's had been unity, and Kepler's was harmony. As to the working out of actual details, Galileo appears to have cared about this only when some practical end was in view, as with his tables of the satellites of Jupiter, which were intended to be used for the determination of longitude at sea, or his tables of trajectories for mortar shots or for other low-speed projectiles. The ultimate value of theories, as Galileo saw it, was expressed thus in the *Dialogue:*

> There is not a single effect in nature, not even the least that exists, such that the most ingenious theorists can ever arrive at a complete understanding of it. . . . Anyone who had experienced just once the perfect understanding of one single thing, and had truly tasted how knowledge is accomplished, would recognize that of the infinity of other truths he understands nothing.[7]

Of the three great Copernicans during the hundred years that ran from the death of Copernicus to the death of Galileo, I think it is fair to say that only Kepler pursued faithfully the exact goals of Copernicus, and Kepler brought the system to perfection. Those goals were astronomical and mathematical, not metaphysical or physical; and to the extent that Kepler tried to add such embellishments, they were not successful. Bruno pursued the supposed implication of an infinite universe, which found favor in the eighteenth and nineteenth centuries but is no longer a part of science. Galileo contributed the beginnings of the physics that ultimately fitted with the new system, but the actual integration of the two had to await Newton's universal gravitation.

Although neither Bruno nor Galileo pursued the principal goal of Copernicanism beyond the point which its author himself had reached, it was their work, rather than Kepler's, that played the decisive role in establishing the place of science in Western culture. For it was not Copernicus or Kepler, but Bruno and Galileo, who made the Copernican theory revolutionary, in the sense in which Professor Kuhn remarked that *De revolutionibus* itself was unrevolutionary. It is on this point that I shall frame my concluding remarks, and for that purpose I shall return briefly to Bruno.

Following his Copernican interlude in England, Bruno returned to France, and thence via Germany to Prague, where he tried to interest the Emperor in a program of religious toleration and benevolent government. From 1588 to 1591 he published works in which can be discerned a profound belief that he had found the one true philosophy under which all the warring elements on earth could find peace and

[7] Galileo, *Dialogue concerning the two chief world systems* . . ., trans. S. Drake, p. 101, Berkeley, 1953.

unity. Then, despite personal dangers that are now obvious to us, the ex-Dominican returned to Italy. The immediate pretext was an invitation by a Venetian nobleman to teach him his philosophy. This turned out to be a trap, and Bruno was soon turned over to the Inquisition; but I am not at all certain that this event was unforeseen by Bruno. From what is left of the proceedings against him, it seems to me quite possible that when Bruno returned to Italy he believed himself able, if only he could present his grand scheme to the Pope himself, to inaugurate a reform of religion and philosophy that would restore the lost unity of the Church and of all medieval life, at a much higher plane.

The results were personally disastrous for Bruno. After a rather surprising surrender of his person by the Venetians to the Roman Inquisition, and a long incarceration at Rome during which he refused to admit the power of that tribunal to determine whether or not his doctrines were heretical, he was refused the right to have them passed on to the Pope as he demanded, and early in 1600 he was sentenced to death at the stake. One of the Cardinals Inquisitors was Robert Bellarmine, who heard his calm and brave reply, "Perhaps you feel more fear in pronouncing this sentence than I do on hearing it." These words were bound to stir the consciences of the men who refused the accused the right of final appeal, and perhaps they affected Bellarmine's later dealings with Galileo.

Copernicanism as such was not an issue in Bruno's trial, except perhaps in his own mind; but science was, if I read the record aright. Bruno's plan was to restore the integral unity of science, philosophy and religion, as that had always existed in times past. The conditions necessary to accomplish this under Bruno's proposed alterations in science and philosophy, inspired by his Copernicanism, were unacceptable to the inquisitioners, and Bruno's program was utterly rejected without its even being submitted to the highest authority.

Galileo's program a few years later was utterly different. Far from proposing to integrate science and religion, Galileo saw them as separate specialities, to be conducted by separate groups of experts, working by different methods. Scientists would declare the facts of nature, and theologians would then explain how these accorded with the truths of religion. The new astronomy could be understood only by experts, whom the Church should consult rather than govern. This theme, found not only in the *Letter to Christina*, but in the important preface to the *Dialogue* (Fig. 67) for which Galileo managed to get the official imprimatur, was in the end also rejected by the Church, though it appears that while Cardinal Bellarmine was still alive he favored its adoption. Only late in the nineteenth century did something very like it become official Catholic policy under the bull *De providentia dei*, after the debates over Darwinism.

Our modern disputes over the relations of philosophy and science have their roots in the widely different Copernicanisms of Bruno and Galileo. Meanwhile, science itself has grown—as did the Copernicanism of Kepler—by patiently modifying what is received from the past to fit more closely the more numerous and more accurate observations of the present. Like Kepler's motto, the motto of science has been: "My witnesses are time and light, and those who hear those witnesses, as will the erudite and upright."[8]

[8] J. Kepler, *Gesammelte Werke*, ed. M. Caspar, vol. XIV, p. 445 (Munich, 1949). (Letter written to Edmund Bruce at Padua.)

Fig. 63. Giordano Bruno (1548–1600). Engraving, seventeenth century. (Courtesy of the Burndy Library.)

Fig. 64. Portrait of Kepler from a contemporary engraving by Jacob von Heyden. (Courtesy of the Burndy Library.)

FIG. 65. Portrait of Galileo by the Flemish painter Justus Sustermans, which hangs in the Uffici in Florence. (Courtesy of the Burndy Library.)

FIG. 66. An eighteenth-century engraving of Galileo from the painting by Justus Sustermans, 1635. (Courtesy of the Burndy Library.)

FIG. 67. Frontispiece of Galileo's *Dialogus*, second edition. (Photograph by Charles Eames.)

Copernicus as a Man and Contributor to the Advancement of Science

EDWARD ROSEN

City University of New York, New York

AS WE gather on this joyous occasion to commemorate the 500th birthday of Nicholas Copernicus, the founder of modern astronomy, we would do well to consider his qualities as a man and to assess his contribution to the advancement of science.

Like other great men, Copernicus has attracted his fair share of would-be belittlers. One such detractor, in a best-selling lopsided work of fiction that is sometimes mistakenly classified as history of science, labeled Copernicus "The Timid Canon". Canon he surely was, but was he timid?

An all too familiar test of a man's timidity or courage is his behavior during a war. It so happens that in the year 1520 the militaristic and aggressive Order of Teutonic Knights invaded Varmia, the homeland of Copernicus. In that year the Knights celebrated New Year's Day by seizing a Varmian town just inside the northeastern frontier. Later that year enemy troops pressed southward toward Olsztyn (Allenstein, in German), "this outer bastion of the whole diocese", where Copernicus had just begun his second term as administrator for the Cathedral Chapter to which he belonged as one of its sixteen canons. In the face of this powerful armed force of Knights menacing his headquarters, how did Copernicus as the local commander react? He instructed a fellow canon to procure provisions, firearms and artillery for his threatened town. The arrival of reinforcements strengthened his position. Olsztyn was held securely until a ceasefire was arranged early in 1521. Copernicus' conduct during these hostilities, I submit, shows him to be a man of courage, not a timid canon, who would have surrendered without preparing to offer any resistance.

Our detractor is singularly uninformed about this war between Varmia and the Teutonic Order. He tells us that Copernicus "preferred to remain . . . in his tower behind the safe walls of Frauenburg" (Frombork, in Polish) on the northern coast, whereas in reality Copernicus was the local commander in charge of the southern front. Our detractor also tells us that the soldiers "did not attack fortified towns". The fact is that Frombork, the principal town of Varmia and the seat of its Cathedral Chapter, was sacked by the troops of the Teutonic Order on January 23, 1520. But so poorly has our detractor done his homework that he is capable of stating that Varmia or "Ermland lay wedged in between the lands of the Polish King and of the Order of Teutonic Knights". Actually Varmia lay wholly within the Polish kingdom, of which it was a part. Hence it is utterly ludicrous for our detractor to say that Varmia "successfully maintained its independence . . . against . . . the Polish King", whose army successfully defended the Polish province Varmia.

Our detractor's woeful ignorance of political history is matched by his unfamiliarity with geography. Thus he associates Frombork with East Prussia, although it belonged to West Prussia. Speaking of the Vistula, he states that "Danzig . . . lay closer to the river's estuary" than Toruń, Copernicus' birthplace. But Toruń lies some 90 miles due south of Danzig, which is right at the river's estuary, just as much as New York is at the estuary of the Hudson River and Washington is on the estuary of the Potomac.

Our detractor pushes his emphasis on Copernicus' fictitious timidity to the ridiculous extreme of charging that the astronomer "remained, through all his life, fearful of venturing, in any sense, outside the walls" of his birthplace Toruń. This gem of pure inaccuracy was polished by a self-styled biographer of Copernicus. Yet the astronomer did venture outside the walls of his native Toruń to study in Krakow; to travel back and forth to Italy twice; to work in the episcopal palace of Lidzbark (Heilsberg); to tour Varmia in the exercise of his official duties; and to journey to Königsberg (now Kaliningrad) to treat a sick man.

Very well, you may say, but all this pertains to physical courage, the unflinching determination to confront the armed foe in combat and the highway robber on the road. But what about moral and intellectual courage? Our detractor insists that "submission to authority . . . to Ptolemy and Aristotle . . . is perhaps the main clue to Copernicus' personality. It undermined his independence of character and his independence of thought, kept him in self-imposed bondage, and singled him out as an austere relic of the Middle Ages."

If so, why have we assembled here today? Is not the reason plain for all to see, who wish to see, the plain truth? Copernicus did not submit to the authority of Aristotle and Ptolemy, whose thinking had reigned supreme without effective challenge for over a thousand years. It was no "austere relic of the Middle Ages", no "pale, sour, insignificant figure", who had the thorough self-confidence, the magnificent courage, to refute and overthrow the key dogma of the Aristotelian–Ptolemaic cosmology, which had denied that the Earth moves and that it is a planet and a heavenly body.

It required intestinal fortitude on the part of Copernicus to proclaim the planetary status of the Earth. For once the Earth is acknowledged to be a heavenly body, what happens to the sacrosanct and hallowed traditional distinction between heaven and earth? We dwellers on the Earth no longer have to earn our passage to heaven after our death by our good conduct during our life, since we are already in heaven at the moment of our birth on Earth, a heavenly body. No wonder Copernicus feared the opposition of the theologians.

Copernicus had the audacity to announce (*Revolutions*, Book IV, Introduction) that the Earth "is in the highest degree akin" to the Moon. For Aristotle and Ptolemy, there could be no degree of kinship between these two bodies. For the two Greek thinkers, the Moon was a celestial body in unceasing motion, whereas the Earth abided in unceasing rest, being incapable of the motion which is the distinguishing mark of a heavenly body.

While the Earth as a whole is incapable of motion, according to Aristotle, bits and pieces of it can move every which way in straight lines. But the Greek philosopher's theory of motion had to be modified by Copernicus. Our detractor pontificates that "it was this almost hypnotic submission to authority which became Copernicus' undoing, both as a man and a scientist". It was precisely Copernicus' rejection of authority that was his making as a scientist. For he saw clearly, and stated unambiguously, that a body changing its position in space could be subject to two motions at the same time, for instance, a rotation and a descent. To quote a provocative passage from his immortal masterpiece, the *Revolutions*, Book I, Chapter 8:

> "circular" subsists with "rectilinear" as "being alive" with "being sick." Surely Aristotle's division of simple motion into three types, away from the middle, toward the middle, and around the middle, will be construed merely as a logical exercise. In like manner we distinguish line, point, and surface, even though one cannot exist without another, and none of them without body.

Copernicus' transformation of Aristotle's theory of motion from a doctrine in physics to a mere exercise in logic is by no means the only example of his steadfast refusal to bow to naked authority as such. Consider, for example, Aristotle's theory of gravitation. According to the Greek philosopher, all heavy bodies in the Universe descended or gravitated toward the center of the Universe, which for him was identical with the center of the Earth. Aristotle's doctrine of a unique center for all gravitation in the Universe was explicitly rejected by Copernicus. He boldly asserted the physical existence of multiple centers of gravitation, one for each heavenly body, including the Earth. On such crucial and momentous scientific issues we recognize in Copernicus, not the "almost hypnotic submission to authority" denounced by our detractor, but a manly, clear-eyed, independent launching of new ideas that produced far-reaching consequences.

Aristotle constructed his entire cosmos out of four sublunary elements, and a fifth celestial element, the so-called *aither*. To this *aither* Aristotle ascribed natural circular motion. Thus he accounted for the rotation of the celestial spheres, and at the same time for the stability of the Earth, which did not partake of the *aither*. Copernicus, on the other hand, being thoroughly convinced that the Earth moves, had to abandon Aristotle's *aither* as the cause of the circular motion in the heavens. So little submissive is he to the authority of Aristotle that the latter's supreme cosmic ingredient disappears from sight. Since Copernicus had to find a cause of celestial motion that would apply also to the obviously non-ethereal Earth, he resorted to the spherical figure shared by the Earth and the other heavenly bodies.

By the same token he discarded the outermost of Aristotle's four sublunary elements. Of the existence of the three inner elements, Copernicus had ample evidence. Earth, water, air, of course. But of Aristotle's fourth sublunary element, fire, a supposedly invisible sphere or spherical belt encircling the air and lying just below the region of the heavens, Copernicus said: "we see this earthly fire, for we behold no other" (*Revolutions*, I, 8).

Let us now assemble Copernicus' proclamation of the planetary status of the Earth; its kinship with the Moon; its triple motion; its non-central position; its non-uniqueness as a center of gravity; the elimination of the *aither* and of elemental fire; and the composition of motions. Does this constellation of assertions and denials emanate from an "almost hypnotic submission to authority", or is it on the contrary the product of a mind which is conscious of a mighty tradition, respectful of its undeniable achievements, aware of its shortcomings, and eager to advance along new lines?

Copernicus lived in the epoch of the great voyages of discovery and exploration that radically altered the course of human history. He was a student at Krakow when Columbus discovered the New World, and at Bologna when Da Gama rounded the Cape of Good Hope. If Copernicus really was a "relic of the Middle Ages", as our detractor characterizes him, then would he not have closed his eyes and ears to the reports of the returning navigators? But he knows about:

> the islands discovered in our time under the rulers of Spain and Portugal, and especially America. . . . On account of its still undisclosed size it is thought to be a second group of inhabited countries (*orbem terrarum*). There are also many other islands, heretofore unknown. So little reason have we to marvel at the existence of antipodes or antichthones. Indeed, geometrical reasoning about the location of America compels us to believe that it is diametrically opposite the Ganges district of India (*Revolutions*, I, 3).

Ptolemy, whose authority in geography matched his eminence in astronomy, had categorically rejected the view that the entire known earth is completely surrounded by water. Yet Copernicus dared to ask "What are the inhabited countries and the mainland itself but an island larger than the others?" Was that question uttered by a dullard cloistered "in self-imposed bondage" to ancient authority, or by a free mind alert to fresh movements of men and ideas?

Our detractor tells us that "posterity never knew for certain whether [Copernicus] really believed in his system or not". Yet a highly influential section of posterity did not have the slightest doubt about the matter. The Roman Catholic Sacred Congregation of the Index officially declared in 1620 that Copernicus "openly discusses the earth's motion while refuting the arguments put forward by the ancients to prove that it remains at rest". Whereas our detractor calls Copernicus' *Revolutions* "The Book That Nobody Read", the Sacred Congregation referred to Copernicus as "the well-known astronomer". As for his sincerity, the Congregation judged that he "has no hesitation in asserting, not as hypothesis but as solid truth, principles concerning the position and motion of the terrestrial globe which are incompatible with Holy Scripture".

Incidentally, if Copernicus' *Revolutions* was indeed "The Book That Nobody Read", one may be permitted to wonder why hard-nosed businesslike publishers invested their money in issuing the first edition at Nuremberg in 1543, the second edition at Basel in 1566, and even more strikingly the third edition at Amsterdam in 1617, when the format was adjusted to a set of contemporary tables, so that the student and practical astronomer could have both works bound conveniently together. If nobody read the *Revolutions*, why was it reprinted at Warsaw in 1854, at Thorn in 1873, at Munich and Berlin in 1949? Why have facsimiles of the first edition been produced over and over again? Why was a photocopy of Copernicus' autograph published at Munich and Berlin in 1944, and again at Warsaw and London in 1972? Why was the *Revolutions* translated into Polish in 1854 and into German in 1879, and why was the German translation reissued in 1939? Why was the English translation included in the *Great Books of the Western World* in 1952? Why did a Russian translation appear in 1964 and a Spanish in 1969? Why are new Polish, German, and English translations about to be published, not to mention the first translation into French? Why is the International Business Machines Corporation (IBM) at this very moment displaying a stunning Copernicus Commemoration in its New York Exhibition Hall of Art and Science? Why is UNESCO preparing for Eurovision a truly spectacular telecast on the occasion of Copernicus' 500th birthday next February 19? Is all this multi-national effort a huge deception? Are we all being hoodwinked into celebrating a "Book That Nobody Read"?

Let us grant that Copernicus' *Revolutions* was never a best-seller, like our detractor's collection of spiteful misinformation. Even if Copernicus' *Revolutions* had acquired only a single reader, the onward march of soundly informed thought about the physical Universe would have progressed without interruption, provided that unique reader was Michael Mästlin, professor of astronomy at Tübingen University. Fortunately, Mästlin's copy of the *Revolutions* still survives. His numerous marginal notations instantly reveal the minute care with which he studied Copernicus' masterpiece. At an early age he became convinced that Copernicus was right, and he lectured on the subject in his university classes. One of his pupils was Johannes Kepler. That immortal genius took the sputtering torch of Copernicanism from Mästlin's cautious hands and, by injecting new fuel, raised high the bright flame by which mankind has ever since been guided toward increasingly profounder penetration of nature's hidden mysteries. The reading by Mästlin of "The Book That Nobody Read" was vastly more momentous for the human race than all the pointless reading of a misconceived best-seller, which calls Mästlin a sacred cow.

The author of that best-seller evidently did not read "The Book That Nobody Read". Had he done so, he would not have referred to Copernicus' "basic star, the Spica". That star, in the constellation of the Virgin, is assigned celestial longitude 170° in the Star Catalogue of Copernicus, whose basic star at 0°0′ longitude, almost diametrically opposite Spica, is, "of the two stars in the horn of the Ram, the one to the west".

Our detractor remarks about the *Revolutions* that "no summary, conclusions, or winding-up of any kind is found at the end of the book, although we are repeatedly promised it in the text". Had our detractor really read the *Revolutions*, he would surely, in one of his 150 footnotes, occupying 19 pages, have told us where to look for this repeated promise, which no reader of the *Revolutions* has ever noticed.

Our detractor assures us that "the name of [Domenico Maria da] Novara [was] omitted". But the name of Copernicus' teacher at the University of Bologna was mentioned in the autograph of the *Revolutions* at folio 79, recto, line 10 up. Moreover, the information that Copernicus "was not so much the pupil as the assistant and witness of observations of the learned Domenico Maria" da Novara was imparted by Copernicus to his disciple Rheticus, who promptly printed that statement in his *First Report*, published in 1540.

Rheticus' name itself was omitted from the *Revolutions*, an omission melodramatically branded by our detractor as a "scandal" and "betrayal of Rheticus". Yet when Rheticus first saw printed copies of the *Revolutions*, he exploded in violent rage about one feature, but did not complain about the omission of his name. His attitude toward Copernicus underwent no change as a result of that omission. On the contrary, Rheticus later declared publicly that Copernicus "is never praised enough" and avowed that he had "always cherished, esteemed, and honored Copernicus not only as a teacher but also as a father". His own father having been beheaded as a supposed sorcerer, Rheticus, a Protestant and Professor of Mathematics at Wittenberg University, the militant think-tank of the anti-papal Lutheran heresy, realized full well that his name would be out of place in a work dedicated to Pope Paul III and honoring the Cardinal of Capua and the Bishop of Chełmno (Kulm).

Our detractor is so well informed about the career of Rheticus that he makes him both a student and a professor at the University of Göttingen, which was inaugurated a century and a half after Rheticus' death. Moreover, our detractor calls Rheticus "an adoring disciple and, fortunately, either homo- or bi-sexual. . . . I say 'fortunately' because the so afflicted have always proved to be the most devoted teachers and disciples, . . . and History owes them a debt."

Let us examine this precious debt. Rheticus had to leave Leipzig University in the spring of 1551 because he had made a male student drunk and had had sexual relations with him. This incident is thoroughly documented in the records of Leipzig University. Our detractor argues that the Leipzig episode "was a repetition of the incidents which, eight years earlier, had caused his migration from Wittenberg to Leipzig". But no such incident is known to have occurred at Wittenberg. Had it been known, would Philip Melanchthon, the stern preceptor of Germany, have recommended Rheticus as enthusiastically as he did to his trusting friends at Leipzig? If our detractor's backward inference from Leipzig to Wittenberg is unwarranted, even less valid is its application to Copernicus. When his sexual conduct was criticized by his personal enemies, their attacks always focused on females, never on a male.

The cause of Rheticus' wrathful assault on the first edition of the *Revolutions* was the unauthorized interpolation of the unsigned Foreword by Osiander, Rheticus' successor as the editor of the *Revolutions*. Our detractor professes to know that this Foreword "was written around November 1542". For this date he offers not a single shred of evidence. He also contends that Copernicus knew about Osiander's Foreword. For this contention he adduces a letter written by Johann Praetorius to Herwart von Hohenburg in 1609 and stating that "the first few pages were sent to Copernicus. Shortly afterward, however, Copernicus died before he could see the whole work" (Zinner, *Entstehung*, p. 454). The second statement is contrary to historical fact, since a printed copy of the *Revolutions* reached Copernicus on his dying day. But "the first few pages" presumably contained the opening signatures or gatherings, say a–b–c, Book I, Chapters 1–11, not the front matter, which was usually printed last. Since Osiander's interpolated Foreword appeared on the verso of the title-page, it did not reach Copernicus. For Praetorius' letter goes on to say: "Rheticus earnestly asserted that Osiander's Foreword would have displeased Copernicus thoroughly, indeed he would have been enraged to no mean degree. And this is probable, because his philosophy was different from what is asserted in that Foreword, as is clear from his Dedication–Preface and from his book."

Osiander's fictionalist philosophy of science, that its foundations need not be true nor probable and may even be false, was rejected by Copernicus. He firmly believed in the physical truth of his system; he did not hypocritically propound a doctrine in which he himself did not believe; and, above all, he was a courageous man and not a coward.

Copernicus and the Impact of Printing

OWEN GINGERICH

Smithsonian Astrophysical Observatory and Harvard College Observatory

DURING the Copernican jubilee year scholars have focused on several critical historical questions. An important one asks why a Copernicus appeared in the early 1500s rather than a century earlier or a century later.

Had the ancient astronomy just then reached the brink of its final collapse? Certainly the predictions of planetary positions did not agree very well with the heavens, but there was little systematic observing to detect the discrepancies or to improve the mechanisms. In Copernicus' treatise there are no new data that could not have been found several centuries before. His own tables furnished but little improvement, and none of this could be attributed to his new heliocentric cosmology.

The few *avant garde* astronomers who were calling for the restoration of astronomy looked back enviously to antiquity as the time when giants such as Ptolemy clearly understood the intricate mechanisms of the heavens; they certainly expected to work within a geocentric framework to tidy up those few areas where uncertainties had arisen with the passing millennium. A mythology popular today imagines that an inadequately simple astronomy of Ptolemy had meanwhile been embroidered into an unworkable complexity. In reality, virtually no one understood the Ptolemaic system well enough to embellish it. Thus, within the astronomical tradition itself, there is no clear reason why a Sun-centered cosmology should have been introduced at that particular time.[1]

[1] These issues are discussed in some detail in my " 'Crisis' versus Aesthetic in the Copernican Revolution" on pp. 85–95 in this volume.

In contrast, an examination of the internal societal milieu shows that the world was particularly ripe for change and for an innovative cosmology. The early 1500s were times of vast changes. Oceanic navigators opened new continents and an age of exploration. Da Vinci and Dürer coupled mathematics with art to capture new harmonies of proportion and perspective. Martin Luther successfully set into motion a reformation of the Church. The fall of Constantinople in 1453 and the consequent westward dispersal of scholars and manuscripts brought a fresh sense of the ancient intellectual heritage to Europe. Innovation became accepted as a desirable goal in itself.[2]

Meanwhile, the explosive spread of printing with movable type beginning in the 1450s fanned the sparks of all these movements, including the reform of astronomy.[3] Without printing, Copernicus would have been deprived of the vast majority of his source materials. Even five decades earlier, he could not easily have found the requisite information that built his *De revolutionibus* into the greatest astronomical treatise of its century. And without printing, his manuscript might have languished, virtually forgotten, on the shelves of the cathedral library. The lack of astronomers full of enthusiasm for heliocentrism in the remainder of the century bears witness to the isolated genius that was Copernicus'.

The fifteenth-century astronomer Regiomontanus (Johann Müller of Königsberg) understood the potential role of printed books in the reform of science even when printing was in its infancy. Regiomontanus embarked on a multi-faceted plan that included instrument building, observations, translations, and the composition of new treatises. His printing shop set up in Nuremberg in the early 1470s was fundamental to his program of disseminating improved astronomical knowledge.[4] His first printed work was a new treatise by his teacher Georg Peuerbach, *Theoricae novae planetarum* (Fig. 68), followed shortly by the astronomical poem of Manilius. His most ambitious work was the *Ephemerides* for 1475–1506, a bulky volume with his own painstaking calculations; this exceedingly rare book is the only one that carries an explicit date, 1474 (Fig. 69). At about the same time he issued more popular calendars for 1474, in German as well as in Latin. (See p. 208.)

The most fascinating piece issued in Regiomontanus' printing office is a broadside advertising the works he hoped to print (Fig. 70).[5] No fewer than forty-five are

[2] See Marie B. Hall, "The spirit of innovation in the sixteenth century", in Owen Gingerich, ed., *The Nature of Scientific Discovery*, Washington, 1974.

[3] See Stillman Drake, "Early science and the printed book: the spread of science beyond the universities", *Renaissance and Reformation*, 6, 43–52 (1970); see also Elizabeth L. Eisenstein, "The advent of printing and the problem of the Renaissance", *Past and Present*, no. 45, 19–89 (1972).

[4] Ernst Zinner, *Leben und Wirken des Joh. Müller von Königsberg*, 2nd ed., Osnabrück, 1968.

[5] Ernst Zinner, "Die wissenschaftlichen Bestrebungen Regiomontans", *Beiträge zur Inkunabelkunde*, N.F. 2, 89–103 (1938). A beautiful color facsimile appears in C. Burger, *Buchhändleranzeigen des XV. Jahrhunderts* (Leipzig, 1907), no. 15. A translation of the list, occasionally unreliable, is found in Barnabas Hughes, *Regiomontanus on Triangles*, pp. 15-16, Madison, 1957.

listed, including outside works such as Ptolemy's *Almagestum*, *Quadripartitum*, and *Musica*, Theon's *Commentaria*, Euclid's *Elementa*, Archimedes' *De sphera et cylindro*, Vitelo's *Perspectiva*, Apollonius' *Conica*, Theodosius' *Sphaera* and Hyginus' *Astronomica*, as well as his own treatises on the Triangle, on the motion of the eighth sphere, on the regular solids, on comets, on the squaring of the circle, an abridged *Almagest*, etc. Unfortunately, this promising endeavor came to a premature end; in 1475 Regiomontanus was called to Rome to advise on calendar reform, and he died there of the plague the following summer at the age of 40. Only nine items, including the broadside catalog, had actually been printed.

The spirit of Regiomontanus' printing program was soon taken over by Erhard Ratdolt, the finest scientific printer of the fifteenth century.[6] Ratdolt left his native city of Augsburg in 1474, and began printing in Venice in 1476; perhaps the two intervening years were spent as a printer with Regiomontanus.[7] In any event, his first production in Venice was a Regiomontanus *Calendarium* (1476) notable in bibliographic circles as the first printed work with a title-page (Fig. 71). Ratdolt issued several annual calendars attributed to Regiomontanus. He repeatedly reprinted the *Ephemerides* (1481, etc.) (Fig. 72) and he first printed Regiomontanus' *Tabulae Directionum* (1490). His editions of Sacrobosco's *Sphaera* (1482, 1485) included reprints of Peuerbach's *Theoricaenovae planetarum* and Regiomontanus' own *Disputationes contra Cremonensis Deliramentia*. Other books from Regiomontanus' prospectus were the first edition of Euclid's *Elementa* (1482), Ptolemy's *Quadripartitum* (1484) and the first illustrated edition of Hyginus' *Poetica astronomica* (1482). Besides maintaining high standards of typography, Ratdolt brought to Italy the use of large printed initials (which had already been adopted by Regiomontanus), and he introduced the first multi-color diagrams (in the Peuerbach of 1485) (Fig. 73). In 1488 Ratdolt accepted the entreaties from Augsburg and returned to his home town. There he concentrated on a distinguished series of liturgical books, although his production included excellent scientific editions.

Although Ratdolt was the outstanding printer of scientific incunabula, he was by no means the only one. Copernicus owned a Ratdolt Euclid (Fig. 74), the 1492–1506 *Ephemerides*, and *Tabulae Directionum*, but his well-thumbed copy of the *Alfonsine Tables* was the second edition, of 1492, printed in Venice by Johannes Hamman. Another Hamman edition owned by Copernicus was the *Epytoma in*

[6] See Gilbert R. Redgrave, *Erhard Ratdolt and His Work at Venice*, London, 1894 (with a list of Ratdolt's Venetian works), and Robert Diehl, "Erhard Ratdolt, ein deutscher Meisterdrucker des 15. und 16. Jahrhunderts", *Philobiblon*, **6**, 91–96, 123–8, 155–61 (Vienna, 1933); both contain numerous excellent reproductions of Ratdolt's printing.

[7] This reasonable speculation was, according to K. Haebler, *Die deutschen Buchdrucker des XV. Jahrhunderts im Auslande*, pp. 106–10, Munich, 1924, first suggested by H. F. Brown, *The Venetian Printing Press*, London, 1891.

Almagestum Ptolemai (1496).[8] This abridgment and commentary by Regiomontanus played a particularly important role in providing Copernicus both with data and with ideas about alternative geometrical mechanisms; it may have been one of the most influential sources of information for the young astronomer, and surely it would not have been available to him except through the burgeoning medium of printing.

The advent of the sixteenth century did not produce any abrupt changes in scientific printing. Nevertheless, as printed books evolved further from the manuscript prototypes, they gradually developed their own character. For example, it became increasingly common to find a book's date and publisher on the first rather than on the last page. This is true of two important astronomical books in Copernicus' library from this time, Johann Stoeffler's *Calendarium Romanum magnum* (Oppenheim, 1518), in which Copernicus systematically recorded his observations of eclipses, and the first edition of Ptolemy's *Almagestum* (Venice, 1515), in which his precise annotations and marginal drawings give evidence of careful study.

At the same time more elaborate productions became possible, such as Stoeffler's *Elucidatio fabricae usuque astrolabii* (1513), which contains not only many handsome woodcuts, but numerous foldout extensions to increase the size of the technical diagrams. Moving discs or volvelles were first used by Regiomontanus himself in his 1474 *Kalender*;[9] in the sixteenth century they became a standard teaching aid in the many editions of Petrus Apianus' *Cosmographia* (1529, etc.), and in Johann Schöner's rare *Aequatorium astronomicum* (1521) the volvelles became an actual paper instrument. The most elaborate volvelle book of the century and one of the triumphs of the bookmaking art of all times was Apianus' *Astronomicum caesareum*, an accurate paper equatorium with imperial detail.[10]

In the 1530s Johannes Petreius of Nuremberg emerged as the leading scientific printer of northern Europe. Unlike Regiomontanus and Ratdolt, his output was not exclusively scientific, and in no year did the majority of his titles concern science.[11] Nevertheless, his most famous editions are Copernicus' *De revolutionibus* (1543) and Cardano's *Ars Magnis* (1545).

[8] A useful list of books owned by, or available to, Copernicus is found in L. Jarzębowski's *Biblioteka Mikolaja Kopernika*, Toruń, 1971. An earlier list, of the Copernican books now found in Sweden, is E. Barwiński, L. Birkenmajer and J. Łos, *Sprawozdanie z Poszukiwań w Szwecyi*, pp. 94–119, Krakow, 1914. The location of Copernicus' own copy of the *Epytoma* is at present unknown, and a systematic examination of copies of this scarce incunabulum offers the exciting prospect that the copy may someday be identified.

[9] This work, including the volvelles, is printed in facsimile by Ernst Zinner in *Der deutsche Kalender des Johannes Regiomontan*, Leipzig, 1937.

[10] See Owen Gingerich, "Apianus' *Astronomicum Caesareum* and its Leipzig facsimile", *Journal for the History of Astronomy* 2, 168–77 (1971).

[11] Joseph C. Shipman, "Johannes Petreius, Nuremberg publisher of scientific works, 1524–1550, with a short-title list of his imprints", in Hellmut Lehmann-Haupt, ed., *Homage to a Bookman: Essays on Manuscripts, Books and Printing written for Hans P. Kraus on his 60th Birthday*, pp. 147–62, Berlin, 1967.

Petreius' first scientific production was the 92-page *Ephemeris pro anno MDXXXII* by Johann Schöner, a Nuremberg astronomer.[12] Schöner, having access to the manuscript legacy of Regiomontanus, began at this time to edit a long series of the earlier astronomer-printer's unpublished works. The initial one printed by Petreius, Regiomontanus' *De triangulis* (1533), was the first great Western trigonometry; Sarton has called it "the greatest synthetic accomplishment of his time".[13] In the ensuing decade Schöner wrote or edited half a dozen more works for Petreius, including his *Tabulae astronomicae* (1536). Meanwhile, Petreius temporarily added Apianus to his stable of authors and, in 1541, Girolamo Cardano.

In 1539, when the young Georg Joachim Rheticus set off from Wittenberg to visit Copernicus in northern Poland, he carried with him five books for his learned teacher. Three were printed by Petreius: the Regiomontanus' *Triangles* (Fig. 75), Apianus' *Instrumentum primi mobilis* (1534) (Fig. 76), and Vitello's *Optics* (1535) edited by Tanstetter and Apianus. Besides these were two first-edition Greek texts, printed in Basel where, under Erasmus' influence, fonts of Greek type were readily available: the Ptolemy *Megales Syntaxeos* published jointly with Theon's *Commentary* (1538), and Euclid's *Elements* (1533).

We can almost envision Rheticus in the role of a publisher's agent bringing seductive gifts to persuade the reluctant Copernicus of the merits of publishing in Germany. His close ties to the Nuremberg circle are further emphasized by the fact that Rheticus dedicated his first report *Narratio prima* (Danzig, 1540) to Johann Schöner. How much the handsome gifts, attractively bound into three white pigskin volumes, influenced Copernicus in addition to Rheticus' own persuasive enthusiasm we shall never know. At least the *De triangulis* interested Copernicus enough for him to add a pair of new leaves with additional theorems into his own trigonometry section of *De revolutionibus*.[14]

As it happened, *De revolutionibus* itself did not become Copernicus' first printed scientific work. After Rheticus returned home to Wittenberg, he excerpted the trigonometry section of Book I for local publication as a small textbook, *De lateribus et angulis triangulorum* (1542) (Fig. 77). This work was small enough so that a local

[12] A complete Schöner bibliography is given by C. H. Coote in Henry Stevens, *Johann Schöner, a Reproduction of His Globe of 1523 Long Lost.* Schöner printed with Petreius regularly until 1544 when he abruptly switched for unknown reasons to the firm of Johannes Montanus and Ulrich Neuber. Although Montanus and Neuber used type styles closely similar to Petreius', they must be considered competitors rather than successors. In some respects the successor to Petreius was Heinrich Petri in Basel, a distant relative who took over the printing of Cardano's works and who printed the second edition of Copernicus' book in 1566. For dates and references on these printers, see Josef Benzing, *Die Buchdrucker des 16. und 17. Jahrhunderts im deutschen Sprachgebiet*, vol. 12 of *Beiträge zum Buch- und Bibliothekwesen*, Wiesbaden, 1963.

[13] George Sarton, "The scientific literature transmitted through the incunabula", *Osiris*, 5, 41–245, esp. p. 68 (1938).

[14] *Nicholas Copernicus Complete Works, I, The Manuscript of Nicholas Copernicus' "On the Revolutions" Facsimile*, Krakow, 1972, folios 24–25.

printer could easily cope with its printing and distribution. In effect, it served as a curtain-raiser to the opus magnum. *De lateribus* was also a harbinger of an important aspect of astronomical printing in the latter half of the century: the inexpensive elementary textbook.[15] Today examples of these once-cheap texts are rarities, but their numerous editions from university centers such as Wittenberg testify to their importance in training astronomers. To the historian of science they furnish remarkable evidence; although Copernicus' name is occasionally mentioned, the heliocentric system did not appear in an astronomy textbook until Kepler's *Epitome astronomiae Copernicanae* (1617–21).[16]

Tantalizingly little information survives concerning the printing history of *De revolutionibus* itself, and none of the technical details are known. We do not know the time required for the printing, the size of the edition, the financial arrangements, the methods of distribution, or the price.[17] Some things can be conjectured from the standard practices of the day. Thus we can deduce that if a single press were used for the folio sheets, the printing time would have been about 4 months, assuming that the woodcut diagrams were ready. It is likely that the type would have been redistributed and continually reused, so that a competent technical proofreader would have been required on the scene. By this time Rheticus had taken a professorship at Leipzig, which was still too far from Nuremberg for him to have assisted directly with the production. Thus Petreius turned to a local scholar, Andreas Osiander, who had previously served as editor for at least one of the Nuremberg printer's productions.

In order to disarm criticism of the unorthodox cosmology in the book, Osiander added an unsigned introduction on the nature of the hypothesis. Rheticus was so offended by the introduction that he struck it out in copies that he distributed. The title-page of one of them is shown in Fig. 78, where another Rheticus deletion, the added words "Orbium Coelestium" in the title, can be seen. The date of his inscription, April 20, 1543, sets a final limit on the date of the printing, at the very least 5 weeks before the death of Copernicus on 24 May, 1543.

Wildly diverse guesses about the size of the first edition have appeared in the literature. At the present time locations of approximately 200 copies are known; perhaps an additional hundred still exist. This suggests an edition of at least 400, and

[15] See Francis R. Johnson, "Astronomical text-books in the sixteenth century", in E. A. Underwood, ed., *Science, Medicine and History*, pp. 285–302, London, 1953.

[16] See Owen Gingerich, "From Copernicus to Kepler: Heliocentrism as model and as reality", *Proceedings of the American Philosophical Society*, 1974 (in press).

[17] The Renaissance bibliophile Achilles Permin Gasser always wrote the purchase price in his books, but unfortunately for the record, Petreius gave him a free copy, which is now in the Vatican Library, Palatine III–103(1). According to Ernst Zinner, *Entstehung und Ausbreitung der Copernicanischen Lehre* (Erlangen, 1943), p. 455, a copy in the Dresden Landesbibliothek bought by V. Engelhardt in Wittenberg in 1545 cost one gulden (very roughly $20). In 1570 Michael Mästlin in Tübingen paid the widow of Victorin Strigel two florins for a copy (again very roughly $20, with depreciation of currency); this is marked inside the back cover of the copy preserved in the Schaffhausen public library.

possibly as many as 800. However, if 800 were sold, it seems improbable that a second edition of about the same size would have been required 23 years later. In any event, enough copies were issued so that its ideas could not easily be suppressed or forgotten.

In the absence of printed sources, it is unlikely that Copernicus could have acquired the technical information that supported his system and gave credibility to his competence as an astronomical reformer. Had his book not been printed, it is unlikely that heliocentric cosmology would have been seriously discussed before 1600—a statement that can easily be justified when we notice how few defenders Copernicus himself had prior to Kepler and Galileo. Thus printing, the primary agent for the transmission of ideas in the sixteenth century, played a fundamental role in the Copernican revolution (Fig. 79).[18]

[18] Reference should also be made to Figs. 80–85 (pp. 215–20), and to the end-papers (Eds.).

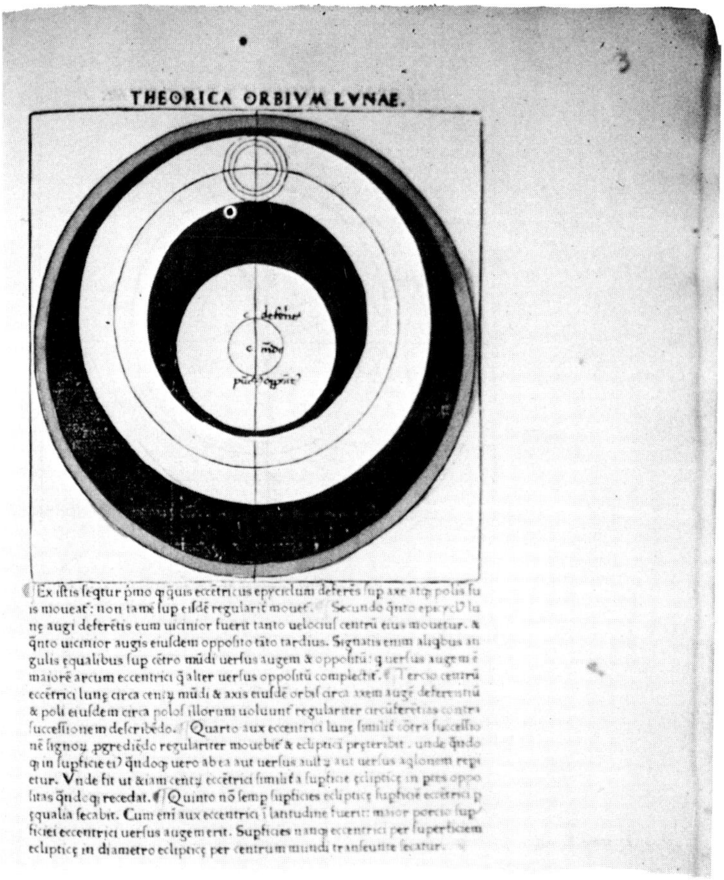

FIG. 68. The Regiomontanus edition of Peuerbach's *Theoricae novae planetarum* (1473). (Courtesy of the Crawford Library, Royal Observatory, Edinburgh.)

FIG. 69. The colophon of Regiomontanus' exceedingly rare *Ephemerides*, his only work that explicitly states the date of printing. (Photograph by Owen Gingerich, courtesy of the Martin-Luther Universitäts Bibliothek, Halle.)

Hęc opera fient in oppido Nuremberga Germanię ductu Ioannis de Monteregio.

ALIENA.

Theoricę nouę planetarum Georgii Purbachii astronomi celebratiſſimi: cum figurationibus oportunis.

Marci Manlii aſtronomica. Hęc duo explicita ſunt.

Coſmographia Ptolemęi noua traductioē. Nā uetula iſta Iacobi Angeli florētini quę uulgo habetur uiciosa ē: interp̄te ipſo (bona ueīa dictū fuerit) neq; līguę grecę ſatis neq; mathēaticę noticiam tenente. Qua ī re ſummis arbitris fidem haberi fas erit: Theodoro Gazę clariſſimo uiro ac grecę latine q; doctiſſimo: & Paulo Florentino grecarū quidem haud ignaro: in mathematicis aūt plurimū excellenti.

Magna compoſitio Ptolemęi: quā uulgo uocāt Almaieſtū noua traductione.

Euclidis elemēta cum anaphoricis Hypſiclis editione Campani euulſis tamen pleriſq; mendis: quę ꝓprio etiā īdicabūtur cōmentariolo.

Theonis alexandrini clariſſimi mathematici commentaria in Almaieſtum.

Procli ſuſformationes aſtronomicę.

Quadriptitum Ptolemęi & Centum fructus eiuſdē noua traductione.

Iulius Firmicus quantus reperitur.

Leopoldus de Auſtria. & ſi qui alii p̄dictores aſtrologici illuſtratione digni uidebuntur. Nam Antonii quoq; de Montulmo q̄uis fragmenta in uſum multiplicem exponentur.

Archimedis geometrę acutiſſimi opa de ſphęra & cylindro. De circuli dimenſione. De conalibus & ſphęralibus. De lineis ſpiralibus. De ęquipōderātibus. De quadratura parabolę. De harenę numero. Cum cōmētariis Eutocii aſcalonitę ī tria opera ex prędictis: ſcilicet de ſphęra & cylindro. de dimenſione circuli. de ęquipōderātibus. Traductio eſt Iacobi cremonēſis: ſed nō nuſq̄ emendata.

Perſpectiua Vitelonis. opus ingens ac nobile.

Perſpectiua Ptolemęi.

Muſica Ptolemęi cum expoſitione Porhyrii.

Menelai ſphęrica noua editione.

Theodoſii ſphęrica. item de habitationibus. de diebus & noctibus noua traductione.

Apollonii pergenſis conica. Item Sereni cylindrica.

Heronis inuenta ſpiritualia. Opus mechanicū mirę uolptatis

Elementa arithmetica Iordani. Data eiuſdem arithmetica.

Quadriptitū numerorū. Opus uariis ſcatens argutiis

Problemata mechanica Ariſtotelis.

Hygini Aſtronomia cum deformatione imaginū cęleſtium.

Facta p̄terea eſt arbor rhetoricę tullianę ſpecioſa imagine.

Et fiet deſcriptio totius habitabilis notę quam uulgo appellant Mappam mundi. Cęterū Germanię particularis tabula: itē Italię: Hiſpanię: Gallię uniuerſę: Gręcięq;. Sed & ſuas cuiq; hiſtorias ex auctoribus plurimis curſim colligere ſtatutū eſt: quę uidelicet ad montes: quę ad maria: ad lacus amnes q; ac alia particularia loca ſpectare uidebuntur.

OPIFICIS tentata. quę eſſent neꝓdenda an non: pudor ingenuus & reſpublica litteraria diu īter ſe diſceptauere. Ratio audendum cenſuit.

Kalendarium nouū quo promūt̄ coniūctiones uerę atq; oppoſitiones lumīarium. item q; eclipſes eorūdē figuratę. loca luminarium uera quotidie. horarū tam ęquinoctialium q̄ temporalium diſcrimiā duplici inſtrumento ad quaſuis habitationes. ac alia pluria ſcitu iucūdiſſima.

Ephemerides quas uulgo uocāt Almanach ad trigintaduos annos futuros. ubi quotidie intueberis ueros motus oīm planetarū capitiſq; draconis lunaris: una cū aſpectibus lunę ad ſolem & planetas. horis etiam aſpectuū eorundem haud friuole adnotatis. neq; planetarū inter ſe aſpectibus pręterm iſſis. In frontibus paginarum poſita ſunt indicia latitudinū. Eclipſes deniq; luminarium ſi quę futurę ſunt locis ſuis effigurantur.

Hęc duo opera iam prope abſoluta ſunt.

Commentaria magna in Coſmographiā Ptolemęi: ubi exponitur fabrica uſuſq; inſtrumēti Meteoroſcopii: quo Ptolemęus ipſe uniuerſos ferme numeros toti⁹ operis ſui elicuit. Falſo enī quiſpiā crediderit tot lōgitudinum latitudinum q; numeros per ſupernorum obſeruationes īnotuiſſe. Pręterea deſcriptio ſphęrę armillaris una cū tota habitabili ī plano ita dilucidatur ut pleriq; omnes diſcere q̄ant. quam nemo antehac latine intellexit uicio traductoris obſtante.

Commentariolū ſingulare contra traductionem Iacobi Angeli Florentini. quod ad arbitros mittetur.

Theonis alexandrini defenſio ī ſex uoluminib⁹ cōtra Georgiū Trapezuntiū. ubi plane quis deprehendet friuola eius eſſe cōmētaria ī Almaieſtū. traductionē q; ipſam operis ptolemaici uicio non carere.

Commentariolum quo cōmonſtrantur placita Campani ex editione elementorum geometricorum reiicienda.

De quinq; corporibus ęquilateris quę uulgo regularia nūcupātur: quę uidelicet eorum locū impleant corporalē & quę nō. contra commentatorē Ariſtotelis Auerroem.

Cōmētaria ī libros Archimedis eos qui Eutocii expoſitiōe carēt

De quadratura circuli. contra Nicolaum Cuſenſem.

De directionibus. contra Archidiaconum parmenſem.

De diſtinctione domiciliorū cęli. contra Campanū & Ioannem Gazulū raguſinū: cuius & alia de horis temporalibus decreta ibidem retractantur

De motu octauę ſphęrę. contra Tebith ſuos q; ſectatores.

De inſtauratione kalendarii eccleſię.

Breuiarium Almaieſti.

De triangulis omnimodis quīq; uolumina.

Problemata aſtronomica ad Almaieſtum totum ſpectantia.

De Cometę magnitudīe remotiōe q; a t̄ra. de loco ei⁹ uero & cęt

Problemata geometrica oīmoda. Opus fructuoſę iucūditatis.

Ludus pānoniēſis. quē alias uocare libuit Tabulas directionū

Tabula magna p̄mi mobilis cū uſu multiplici ratōnib⁹ q; certis

Radii uiſorii multorum generum cū uſibus ſuis.

De pōderib⁹ et aquęductib⁹ cū figuratōnib⁹ īſtrumētoꝝ ad eas res neceſſariorum.

De ſpeculis uſtoriis atq; aliis multorū generū uſuſq; ſtupendi.

In officina fabrili Aſtrarium in continuo tractatu eſt. Opus plane pro miraculo ſpectandum. Fiunt & alia inſtrumenta aſtronomica ad obſeruationes cęleſtium. item q; alia ad uſum uulgarem quotidianum. quorum nomina longum eſt recitare.

Poſtremo omnium artem illam mirificam litterarum formatricem monimentis ſtabilibus mādare decretum eſt. (deus bone faueas) qua re explicita ſi mox obdormierit opifex mors acerba non erit: quom tantum munus poſteris in hęreditate reliquerit: quo ipſi ſe ab inopia librorum ꝑpetuo poterunt uindicare.

FIG. 70. Regiomontanus' trade list (1474) of books he planned to publish. Only two examples are known, one at the British Museum, the other at the Staatsbibliothek in Munich. (Reproduced from C. Burger, *Buchhändleranzeigen* (1907), by permission of Harvard College Library.)

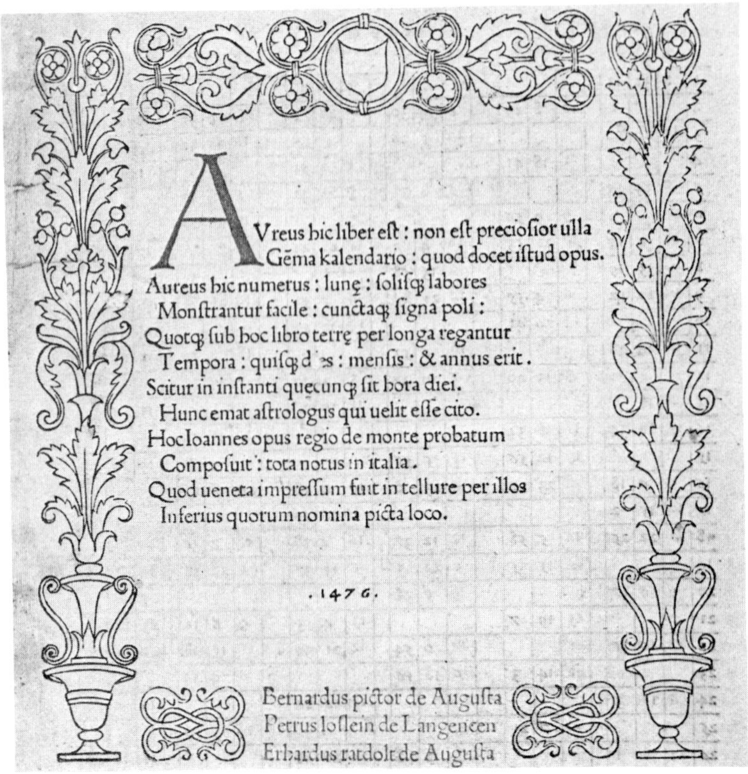

FIG. 71. The first printed title-page on the Regiomontanus calendar of 1476, printed by Ratdolt and his partners in Venice. (Courtesy of the Burndy Library.)

FIG. 72. Part of the corrections for Jupiter in the Ratdolt *Alfonsine Tables* (Venice, 1483). (Collection of Harrison Horblit.)

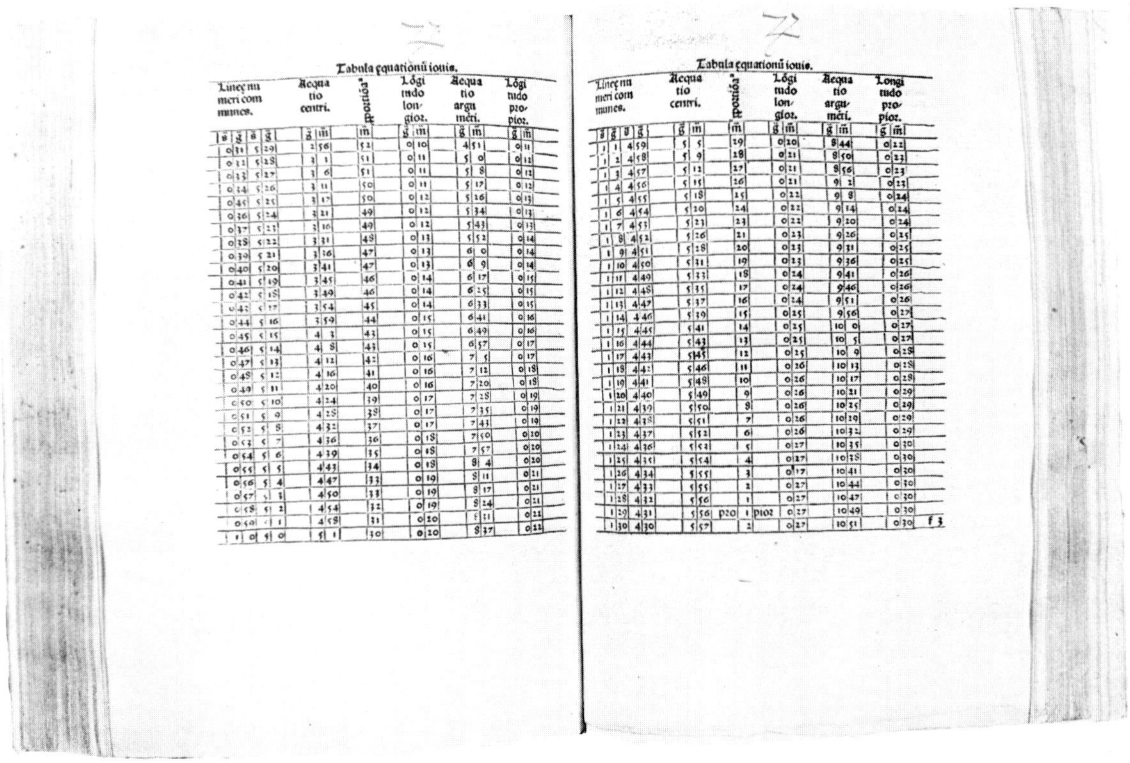

FIG. 73. Large initial and colored diagram, typical of Ratdolt's printing from his 1485 edition of Peuerbach's *Theoricae novae planetarum*. (By permission of the Harvard College Library.)

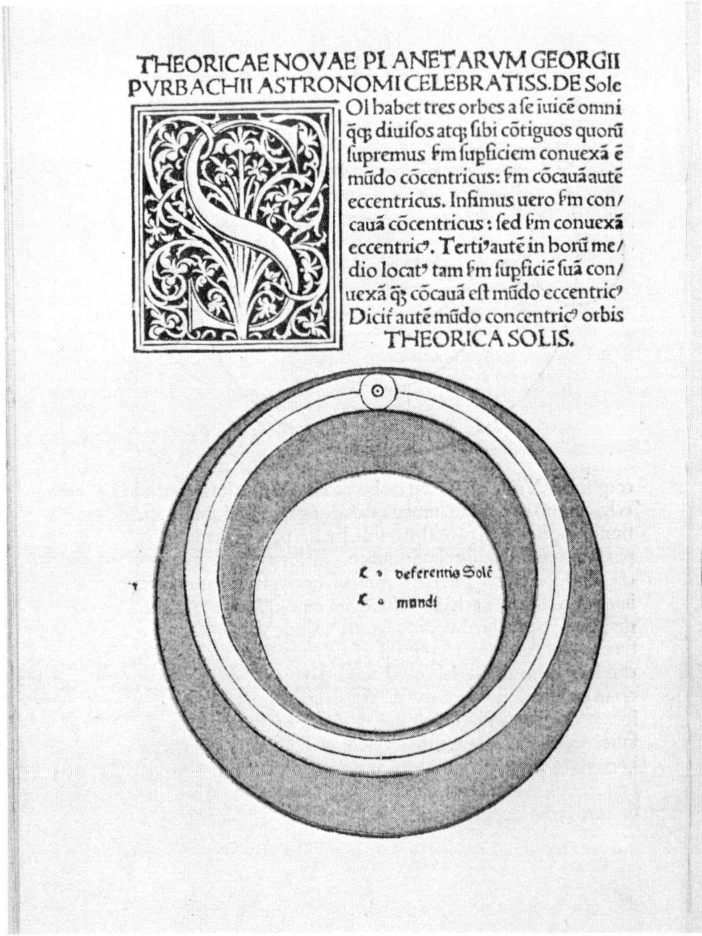

FIG. 74. A broadside advertisement for Ratdolt's *editio princeps* of Euclid (1483). This unique copy was preserved in a bookbinding. (Reproduced from C. Burger, *Buchhändleranzeigen*, 1907, by permission of Harvard College Library.)

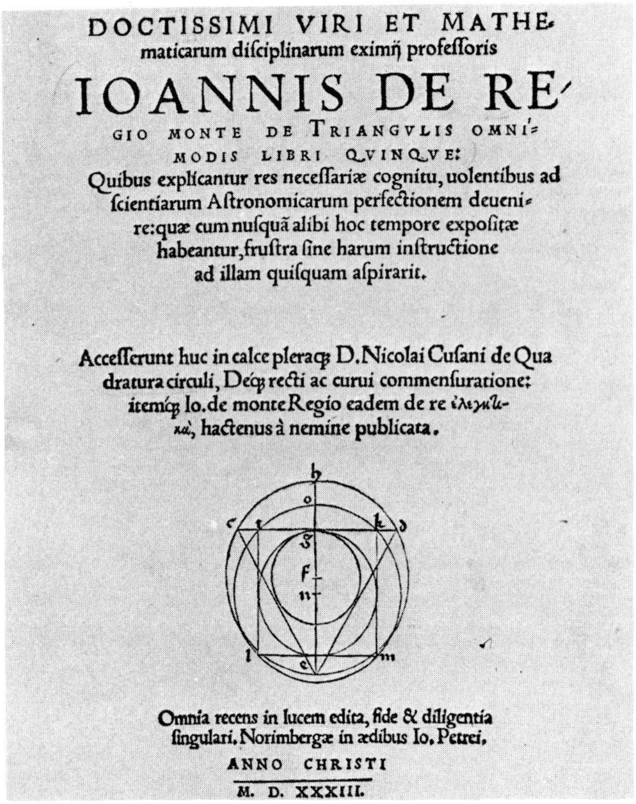

DOCTISSIMI VIRI ET MATHE-
maticarum diſciplinarum eximij profeſſoris
IOANNIS DE RE-
GIO MONTE DE TRIANGVLIS OMNI-
MODIS LIBRI QVINQVE:
Quibus explicantur res neceſſariæ cognitu, uolentibus ad
ſcientiarum Aſtronomicarum perfectionem deueni-
re: quæ cum nuſquã alibi hoc tempore expoſitæ
habeantur, fruſtra ſine harum inſtructione
ad illam quiſquam aſpirarit.

Acceſſerunt huc in calce pleraq; D. Nicolai Cuſani de Qua
dratura circuli, Deq; recti ac curui commenſuratione:
itemq; Io. de monte Regio eadem de re ἐλεγκτικά, hactenus à nemine publicata.

Omnia recens in lucem edita, fide & diligentia
ſingulari. Norimbergæ in ædibus Io. Petrei.
ANNO CHRISTI
M. D. XXXIII.

FIG. 75. The Petreius edition of Regiomontanus' *Triangles* (1533), with a short abstract of the contents on the title, typical of this printer. (Private collection of R. B. and M. S. Honeyman.)

INSTRVMENTVM
PRIMI MOBILIS, A' PETRO APIANO
NVNC PRIMVM ET INVENTVM ET IN LVCEM EDITVM.
Ad cuius declarationem & intellectũ Pronunciata centũ hic proponuntur, è quibus Inſtrumenti nobiliſſimi uſus innoteſcit & compoſitio. Inquirere autẽ & inuenire licebit in hoc inſtrumento, quicquid uſpiam in uniuerſo primo mobili noua quadã ſinuum ratione indagari poteſt: nec quicquã in eo ipſo primo mobili deſiderare poterit, quod nõ per inſtrumentum hoc inueniri facile queat.
Accedunt ijs
GEBRI FILII AFFLA HISPALENSIS ASTRO
nomi uetuſtiſſimi pariter & peritiſſimi, libri IX. de Aſtronomia, ante aliquot ſecula Arabice ſcripti, & per Girardũ Cremonenſem latinitate donati,
nunc uero omniũ primum in lucem editi.
Omnia hæc induſtria & beneuolentia Petri Apiani Mathematici prelo cõmiſſa, & Reuerendiſs. in Chriſto patri & D. D. CHRISTOPHORO A' STADIO, &c. ornatiſſimo Præſuli Auguſtenſi, ob illuſtrationem ſuæ familiæ inſignium, dedicata: Quibus & tu ſtudioſe lector benignus fruere, tanto Præſidi perpetuò gratiſſimus.
Dixit Deus:
Fiant luminaria in firma
mento cœli, ut diuidat diem ac
noctem, & ſint IN SIGNA &
tempora, & dies & annos,
ut luceant in firmamẽto
cœli, & illuminent
terram.
Gen. I.
NORIMBERGAE APVD IO. PETREIVM. ANNO M. D. XXXIIII.

FIG. 76. Rheticus' presentation copy to Copernicus of Apianus' *Instrumentum primi mobilis* (Petreius, Nuremberg, 1534). (Photograph by Charles Eames, courtesy of the Uppsala University Library.)

FIG. 77. The title-page of Copernicus' first printed astronomical work, *De lateribus*, published at the instigation of Rheticus in Wittenberg in 1542. (Collection of R. B. and M. S. Honeyman. Photograph by Charles Eames.)

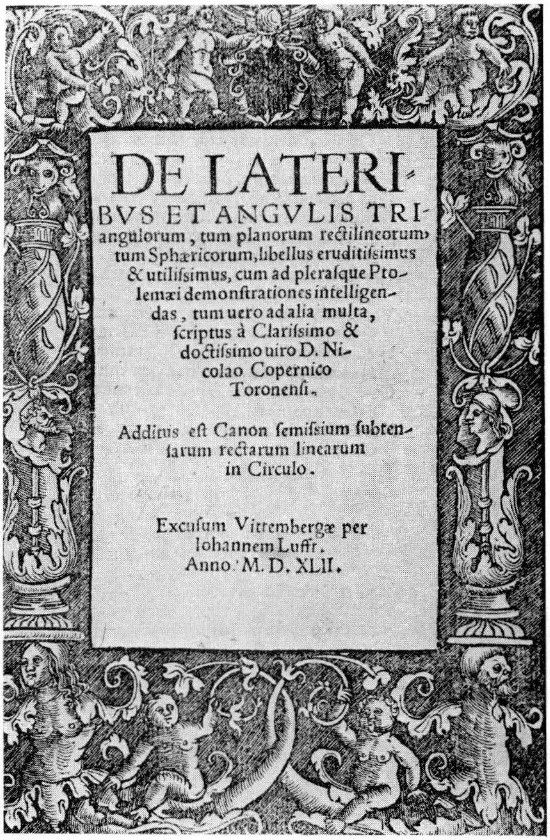

DE LATERIBVS ET ANGVLIS TRIangulorum, tum planorum rectilineorum, tum Sphæricorum, libellus eruditissimus & utilissimus, cum ad plerasque Ptolemæi demonstrationes intelligendas, tum uero ad alia multa, scriptus à Clarissimo & doctissimo uiro D. Nicolao Copernico Toronensi.

Additus est Canon semissium subtensarum rectarum linearum in Circulo.

Excusum Vittembergæ per Iohannem Lufft. Anno M. D. XLII.

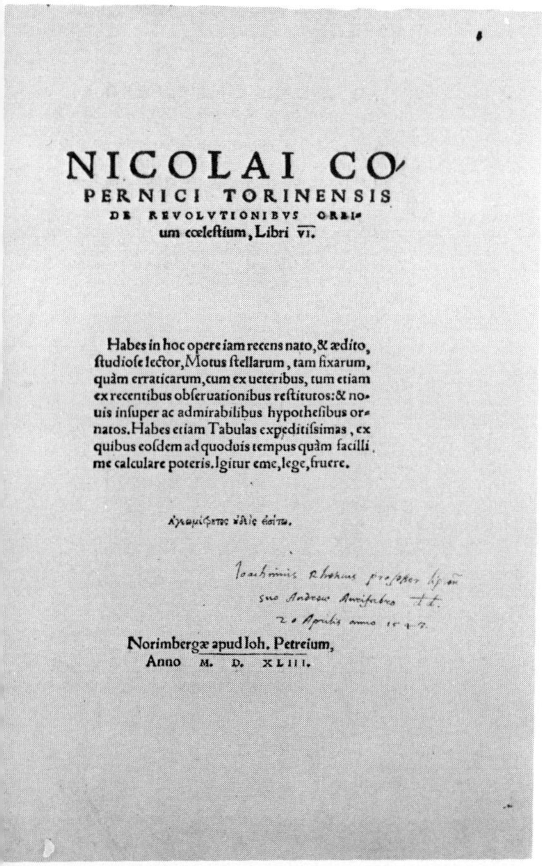

NICOLAI COPERNICI TORINENSIS DE REVOLVTIONIBVS ORBIum cœlestium, Libri VI.

Habes in hoc opere iam recens nato, & ædito, studiose lector, Motus stellarum, tam fixarum, quàm erraticarum, cum ex ueteribus, tum etiam ex recentibus obseruationibus restitutos: & nouis insuper ac admirabilibus hypothesibus ornatos. Habes etiam Tabulas expeditissimas, ex quibus eosdem ad quoduis tempus quàm facillime calculare poteris. Igitur eme, lege, fruere.

Ἀγεωμέτρητος οὐδεὶς εἰσίτω.

Norimbergæ apud Ioh. Petreium, Anno M. D. XLIII.

FIG. 78. Presentation copy of Copernicus' *De revolutionibus* from Joachim Rheticus to Andreas Aurifaber, who was then Dean of the University of Wittenberg. (Photograph by Owen Gingerich, by permission of Harrison Horblit.)

FIG. 79. Books owned or used by Copernicus, now preserved in Uppsala. In the center is Stoeffler's *Calendarium Romanum magnum* (1518) with Copernicus' annotations. In the foreground is Ptolemy's *Almagestum* (1515). Because Copernicus' own copy of the Regiomontanus *Epytoma* is now unknown, a different example is shown at the far left. (Photograph by Owen Gingerich, by permission of the Uppsala University Library and Observatory.)

FIG. 80. Frontispiece to G. B. Riccioli's *Almagestum Novum* (Bologna, 1651): the Copernican System is tested against the Tychonian System and found wanting—a reflection of Riccioli's views. The figure holding the balance is Urania; at her feet lies the rejected Ptolemaic System. Some of the new telescopic discoveries—Jupiter's satellites, Saturn's rings and the Moon's cratered surface—are featured in the upper right of the engraving. (Photograph by courtesy of Prof. Olaf Pedersen, University of Aarhus, Denmark.)

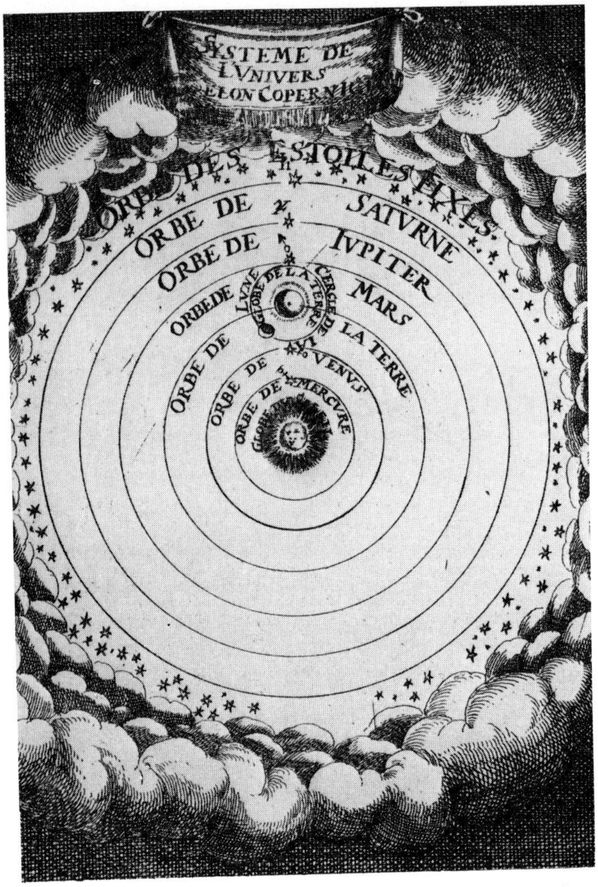

FIG. 81. "Système de Copernic"—from Alain Maresson Mallet's *Description de l'Univers* of 1683 (Vol. I, p. 67, fig. XXII), a full representation of a new World System. (Radio Times Hulton Picture Library, London.)

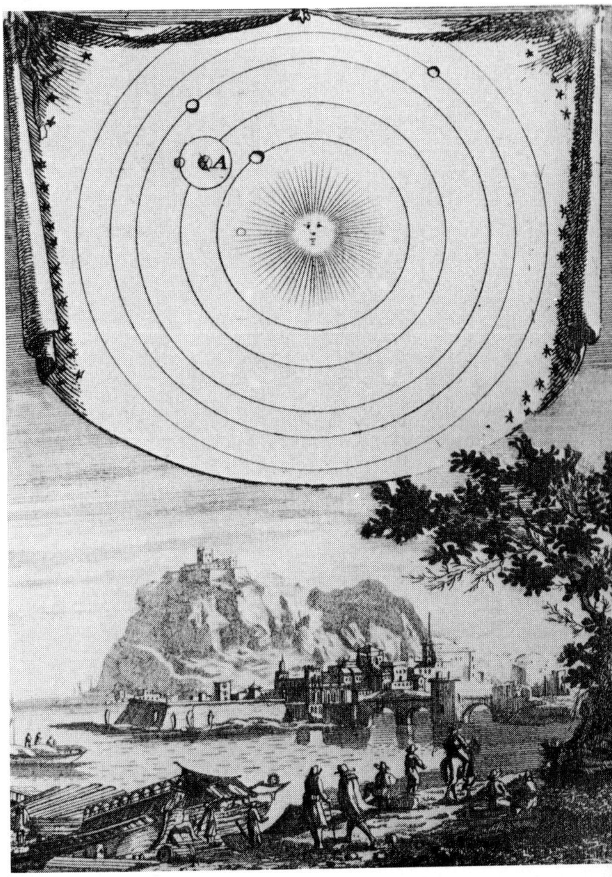

FIG. 82. Another picture from Maresson Mallet's book of 1683, *Description de l'Univers* (Vol. I, p. 73, fig. XXV). It is described as giving "Copernicus' explanation of the 'first movement' of the Earth, i.e. the Earth's diurnal rotation on its axis, which we see as the rotation of the sky in 24 hours". The accompanying landscape is not identified. (Radio Times Hulton Picture Library, London.)

FIG. 83. View of the Copernicus Exhibition at the IBM Exhibit Center in New York. (Courtesy of Charles Eames.)

FIG. 84. Another view of the Copernicus Exhibition at the IBM Exhibit Center in New York. It shows a working model that demonstrates how two systems—the Ptolemaic and the Copernican—can account differently for the same observed phenomena. (Courtesy of Charles Eames.)

FIG. 85. Detail from Hartmann Schedel, *Nuremberg Chronicle*, 1493. (Photograph by Owen Gingerich. Courtesy of Charles Eames.)

Index